COLONEL JACK'S FREE BLOW-OUT

ROUGHING IT

BY
MARK TWAIN
(SAMUEL L. CLEMENS)

IN TWO VOLUMES
VOL. II

HARPER & BROTHERS PUBLISHERS
NEW YORK AND LONDON

ILLUSTRATIONS

CONTENTS

CHAPTER I.

CHAPTER II.

CHAPTER III.

CHAPTER IV.

CHAPTER V.

CHAPTER VI.

CHAPTER VII.

CHAPTER VIII.

CHAPTER IX.

CHAPTER X.

CHAPTER XI.

Contents

CHAPTER XII.

CHAPTER XIII.

CHAPTER XIV.

CHAPTER XV.

CHAPTER XVI.

CHAPTER XVII.

CHAPTER XVIII.

CHAPTER XIX.

CHAPTER XX.

CHAPTER XXI.

CHAPTER XXII.

CHAPTER XXIII.

CHAPTER XXIV.

Contents

CHAPTER XXV.

CHAPTER XXVI.

CHAPTER XXVII.

CHAPTER XXVIII.

CHAPTER XXIX.

CHAPTER XXX.

CHAPTER XXXI.

Contents

APPENDIX

ROUGHING IT

CHAPTER I.

WHAT to do next?

It was a momentous question. I had gone out into the world to shift for myself, at the age of thirteen (for my father had endorsed for friends, and although he left us a sumptuous legacy of pride in his fine Virginian stock and its national distinction, I presently found that I could not live on that alone without occasional bread to wash it down with). I had gained a livelihood in various vocations, but had not dazzled anybody with my successes; still the list was before me, and the amplest liberty in the matter of choosing, provided I wanted to work—which I did not, after being so wealthy. I had once been a grocery clerk, for one day, but had consumed so much sugar in that time that I was relieved from further duty by the proprietor; said he wanted me outside, so that he could have my custom. I had studied law an entire week, and then given it up because it was so prosy and tiresome. I had engaged briefly in the study of blacksmithing, but wasted so much time trying to fix the bellows so that it would blow itself, that the master turned my

2

(15)

adrift in disgrace, and told me I would come to no good. I had been a bookseller's clerk for awhile, but the customers bothered me so much I could not read with any comfort, and so the proprietor gave me a furlough and forgot to put a limit to it. I had clerked in a drug store part of a summer, but my prescriptions were unlucky, and we appeared to sell more stomach pumps than soda water. So I had to go. I had made of myself a tolerable printer, under the impression that I would be another Franklin some day, but somehow had missed the connection thus far. There was no berth open in the Esmeralda *Union*, and besides I had always been such a slow compositor that I looked with envy upon the achievements of apprentices of two years' standing; and when I took a "take," foremen were in the habit of suggesting that it would be wanted "some time during the year." I was a good average St. Louis and New Orleans pilot and by no means ashamed of my abilities in that line; wages were two hundred and fifty dollars a month and no board to pay, and I did long to stand behind a wheel again and never roam any more — but I had been making such an ass of myself lately in grandiloquent letters home about my blind lead and my European excursion that I did what many and many a poor disappointed miner had done before; said "It is all over with me now, and I will never go back home to be pitied — and snubbed." I had been a private secretary, a silver miner and a silver mill operative, and amounted to less than nothing in each, and now—

What to do next?

I yielded to Higbie's appeals and consented to try the mining once more. We climbed far up on the mountain side and went to work on a little rubbishy claim of ours that had a shaft on it eight feet deep. Higbie descended into it and worked bravely with his pick till he had loosened up a deal of rock and dirt, and then I went down with a long-handled shovel (the most awkward invention yet contrived by man) to throw it out. You must brace the shovel forward with the side of your knee till it is full, and then, with a skillful toss, throw it backward over your left shoulder. I made the toss, and landed the mess just on the edge of the shaft and it all came back on my head and down the back of my neck. I never said a word, but climbed out and walked home. I inwardly resolved that I would starve before I would make a target of myself and shoot rubbish at it with a long-handled shovel. I sat down, in the cabin, and gave myself up to solid misery—so to speak. Now in pleasanter days I had amused myself with writing letters to the chief paper of the Territory, the Virginia *Daily Territorial Enterprise*, and had always been surprised when they appeared in print. My good opinion of the editors had steadily declined; for it seemed to me that they might have found something better to fill up with than my literature. I had found a letter in

2*

the post-office as I came home from the hillside,
and finally I opened it. Eureka! [I never did
know what Eureka meant, but it seems to be as
proper a word to heave in as any when no other that
sounds pretty offers.] It was a deliberate offer to
me of Twenty-five Dollars a week to come up to
Virginia and be city editor of the *Enterprise*.

I would have challenged the publisher in the
" blind lead " days — I wanted to fall down and
worship him, now. Twenty-five Dollars a week —
it looked like bloated luxury — a fortune, a sinful
and lavish waste of money. But my transports
cooled when I thought of my inexperience and con-
sequent unfitness for the position — and straightway, on top of this, my long array of failures rose
up before me. Yet if I refused this place I must
presently become dependent upon somebody for my
bread, a thing necessarily distasteful to a man who
had never experienced such a humiliation since he
was thirteen years old. Not much to be proud of,
since it is so common — but then it was all I had to
be proud of. So I was scared into being a city
editor. I would have declined, otherwise. Neces-
sity is the mother of " taking chances." I do not
doubt that if, at that time, I had been offered a
salary to translate the Talmud from the original
Hebrew, I would have accepted — albeit with diffi-
dence and some misgivings — and thrown as much
variety into it as I could for the money.

I went up to Virginia and entered upon my new

vocation. I was a rusty-looking city editor, I am
free to confess — coatless, slouch hat, blue woolen
shirt, pantaloons stuffed into boot-tops, whiskered
half down to the waist, and the universal navy revolver slung to my belt. But I secured a more
Christian costume and discarded the revolver. I
had never had occasion to kill anybody, nor ever
felt a desire to do so, but had worn the thing in
deference to popular sentiment, and in order that I
might not, by its absence, be offensively conspicuous, and a subject of remark. But the other
editors, and all the printers, carried revolvers. I
asked the chief editor and proprietor (Mr. Goodman, I will call him, since it describes him as well as
any name could do) for some instructions with regard to my duties, and he told me to go all over
town and ask all sorts of people all sorts of questions, make notes of the information gained, and
write them out for publication. And he added:

"Never say 'We learn' so-and-so, or 'It is re-
ported, or 'It is rumored,' or 'We understand' so-
and-so, but go to headquarters and get the absolute
facts, and then speak out and say 'It *is* so-and-so.'
Otherwise, people will not put confidence in your
news. Unassailable certainty is the thing that gives
a newspaper the firmest and most valuable reputation."

It was the whole thing in a nutshell; and to this
day, when I find a reporter commencing his article
with "We understand," I gather a suspicion that he

B ﹡﹡

has not taken as much pains to inform himself as he ought to have done. I moralize well, but I did not always practice well when I was a city editor; I let fancy get the upper hand of fact too often when there was a dearth of news. I can never forget my first day's experience as a reporter. I wandered about town questioning everybody, boring everybody, and finding out that nobody knew anything. At the end of five hours my notebook was still barren. I spoke to Mr. Goodman. He said:

"Dan used to make a good thing out of the hay wagons in a dry time when there were no fires or inquests. Are there no hay wagons in from the Truckee? If there are, you might speak of the renewed activity and all that sort of thing, in the hay business, you know. It isn't sensational or exciting, but it fills up and looks business-like."

I canvassed the city again and found one wretched old hay truck dragging in from the country. But I made affluent use of it. I multiplied it by sixteen, brought it into town from sixteen different directions, made sixteen separate items out of it, and got up such another sweat about hay as Virginia City had never seen in the world before.

This was encouraging. Two nonpareil columns had to be filled, and I was getting along. Presently, when things began to look dismal again, a desperado killed a man in a saloon and joy returned once more. I never was so glad over any mere trifle before in my life. I said to the murderer:

" Sir, you are a stranger to me, but you have done me a kindness this day which I can never forget. If whole years of gratitude can be to you any slight compensation, they shall be yours. I was in trouble and you have relieved me nobly and at a time when all seemed dark and drear. Count me your friend from this time forth, for I am not a man to forget a favor.''

If I did not really say that to him I at least felt a sort of itching desire to do it. I wrote up the murder with a hungry attention to details, and when it was finished experienced but one regret — namely, that they had not hanged my benefactor on the spot, so that I could work him up too.

Next I discovered some emigrant wagons going into camp on the plaza and found that they had lately come through the hostile Indian country and had fared rather roughly. I made the best of the item that the circumstances permitted, and felt that if I were not confined within rigid limits by the presence of the reporters of the other papers I could add particulars that would make the article much more interesting. However, I found one wagon that was going on to California, and made some judicious inquiries of the proprietor. When I learned, through his short and surly answers to my cross-questioning, that he was certainly going on and would not be in the city next day to make trouble, I got ahead of the other papers, for I took down his list of names and added his party to the killed and wounded. Having more scope here, I put this wagon through an Indian fight that to this day has no parallel in history.

My two columns were filled. When I read them over in the morning I felt that I had found my legitimate occupation at last. I reasoned within myself that news, and stirring news, too, was what a paper needed, and I felt that I was peculiarly endowed with the ability to furnish it. Mr. Goodman said that I was as good a reporter as Dan. I desired no higher commendation. With encouragement like that, I felt that I could take my pen and murder all the immigrants on the plains if need be, and the interests of the paper demanded it.

CHAPTER II.

HOWEVER, as I grew better acquainted with the business and learned the run of the sources of information I ceased to require the aid of fancy to any large extent, and became able to fill my columns without diverging noticeably from the domain of fact.

I struck up friendships with the reporters of the other journals, and we swapped "regulars" with each other and thus economized work. "Regulars" are permanent sources of news, like courts, bullion returns, "clean-ups" at the quartz mills, and in- quests. Inasmuch as everybody went armed, we had an inquest about every day, and so this department was naturally set down among the "regulars." We had lively papers in those days. My great com- petitor among the reporters was Boggs of the *Union*. He was an excellent reporter. Once in three or four months he would get a little intoxicated, but as a general thing he was a wary and cautious drinker although always ready to tamper a little with the enemy. He had the advantage of me in one thing; he could get the monthly public school report and I could not, because the principal hated the *Enterprise*.

One snowy night when the report was due, I
started out sadly wondering how I was going to get
it. Presently, a few steps up the almost deserted
street I stumbled on Boggs and asked him where
he was going.

"After the school report."

" I'll go along with you."

" No, *sir*. I'll excuse you."

" Just as you say."

A saloon-keeper's boy passed by with a steaming
pitcher of hot punch, and Boggs snuffed the
fragrance gracefully. He gazed fondly after the
boy and saw him start up the *Enterprise* stairs. I
said:

" I wish you could help me get that school business, but since you can't, I must run up to the
Union office and see if I can get them to let me have
a proof of it after they have set it up, though I
don't begin to suppose they will. Good night."

" Hold on a minute. I don't mind getting the
report and sitting around with the boys a little, while
you copy it, if you're willing to drop down to the
principal's with me."

" Now you talk like a rational being. Come
along."

We ploughed a couple of blocks through the
snow, got the report and returned to our office. It
was a short document and soon copied. Meantime
Boggs helped himself to the punch. I gave the
manuscript back to him and we started out to get an

AN INFERIOR SORT OF A MURDER

inquest, for we heard pistol shots near by. We got the particulars with little loss of time, for it was only an inferior sort of bar-room murder, and of little interest to the public, and then we separated. Away at three o'clock in the morning, when we had gone to press and were having a relaxing concert as usual — for some of the printers were good singers and others good performers on the guitar and on that atrocity the accordeon — the proprietor of the *Union* strode in and desired to know if anybody had heard anything of Boggs or the school report. We stated the case, and all turned out to help hunt for the delinquent. We found him standing on a table in a saloon, with an old tin lantern in one hand and the school report in the other, haranguing a gang of intoxicated Cornish miners on the iniquity of squandering the public moneys on education " when hundreds and hundreds of honest hard-working men are literally starving for whisky." [Riotous applause.] He had been assisting in a regal spree with those parties for hours. We dragged him away and put him to bed.

Of course there was no school report in the *Union*, and Boggs held me accountable, though I was innocent of any intention or desire to compass its absence from that paper and was as sorry as any one that the misfortune had occurred.

But we were perfectly friendly. The day that the school report was next due, the proprietor of the " Genesee " mine furnished us a buggy and asked

us to go down and write something about the property — a very common request and one always gladly
acceded to when people furnished buggies, for we
were as fond of pleasure excursions as other people.
In due time we arrived at the " mine " — nothing
but a hole in the ground ninety feet deep, and no
way of getting down into it but by holding on to a
rope and being lowered with a windlass. The workmen had just gone off somewhere to dinner. I was
not strong enough to lower Boggs's bulk; so I took
an unlighted candle in my teeth, made a loop for
my foot in the end of the rope, implored Boggs not
to go to sleep or let the windlass get the start of
him, and then swung out over the shaft. I reached
the bottom muddy and bruised about the elbows,
but safe. I lit the candle, made an examination of
the rock, selected some specimens and shouted to
Boggs to hoist away. No answer. Presently a
head appeared in the circle of daylight away aloft,
and a voice came down:

"Are you all set?"

"All set — hoist away."

"Are you comfortable?"

" Perfectly."

" Could you wait a little?"

" Oh certainly — no particular hurry."

" Well — good by."

" Why?   Where are you going?"

"After the school report!"

And he did.   I staid down there an hour, and sur-

prised the workmen when they hauled up and found
a man on the rope instead of a bucket of rock. I
walked home, too — five miles — up hill. We had
no school report next morning; but the *Union* had.

Six months after my entry into journalism the
grand "flush times" of Silverland began, and they
continued with unabated splendor for three years.
All difficulty about filling up the "local department" ceased, and the only trouble now was how to
make the lengthened columns hold the world of inci-
dents and happenings that came to our literary net
every day. Virginia had grown to be the "livest"
town, for its age and population, that America had
ever produced. The sidewalks swarmed with people
— to such an extent, indeed, that it was generally no
easy matter to stem the human tide. The streets
themselves were just as crowded with quartz wagons,
freight teams, and other vehicles. The procession
was endless. So great was the pack, that buggies
frequently had to wait half an hour for an opportunity to cross the principal street. Joy sat on every
countenance, and there was a glad, almost fierce, in-
tensity in every eye, that told of the money-getting
schemes that were seething in every brain and the
high hope that held sway in every heart. Money
was as plenty as dust; every individual considered
himself wealthy, and a melancholy countenance was
nowhere to be seen. There were military companies,
fire companies, brass bands, banks, hotels, theaters,
"hurdy-gurdy houses," wide-open gambling palaces,

political pow-wows, civic processions, street fights, murders, inquests, riots, a whisky mill every fifteen steps, a Board of Aldermen, a Mayor, a City Surveyor, a City Engineer, a Chief of the Fire Department, with First, Second, and Third Assistants, a Chief of Police, City Marshal, and a large police force, two Boards of Mining Brokers, a dozen breweries, and half a dozen jails and station-houses in full operation, and some talk of building a church. The "flush times" were in magnificent flower! Large fire-proof brick buildings were going up in the principal streets, and the wooden suburbs were spreading out in all directions. Town lots soared up to prices that were amazing.

The great "Comstock lode" stretched its opulent length straight through the town from north to south, and every mine on it was in diligent process of development. One of these mines alone employed six hundred and seventy-five men, and in the matter of elections the adage was, "as the 'Gould & Curry' goes, so goes the city." Laboring men's wages were four and six dollars a day, and they worked in three "shifts" or gangs, and the blasting and picking and shoveling went on without ceasing, night and day.

The "city" of Virginia roosted royally midway up the steep side of Mount Davidson, seven thousand two hundred feet above the level of the sea, and in the clear Nevada atmosphere was visible from a distance of fifty miles! It claimed a population of fifteen thousand to eighteen thousand, and all day long half of this little army swarmed the streets like bees and the other half swarmed among the drifts and tunnels of the "Comstock," hundreds of feet down in the earth directly under those same streets. Often we felt our chairs jar, and heard the faint boom of a blast down in the bowels of the earth under the office.

The mountain side was so steep that the entire town had a slant to it like a roof. Each street was a terrace, and from each to the next street below the descent was forty or fifty feet. The fronts of the houses were level with the street they faced, but their rear first floors were propped on lofty stilts; a man could stand at a rear first-floor window of a C street house and look down the chimneys of the row of houses below him facing D street. It was a laborious climb, in that thin atmosphere, to ascend from D to A street, and you were panting and out of breath when you got there; but you could turn around and go down again like a house a-fire — so to speak. The atmosphere was so rarefied, on account of the great altitude, that one's blood lay near the surface always, and the scratch of a pin was a disaster worth worrying about, for the chances were that a grievous erysipelas would ensue. But to offset this, the thin atmosphere seemed to carry healing to gunshot wounds, and, therefore, to simply shoot your adversary through both lungs was a thing not likely to afford you any permanent satisfaction,

for he would be nearly certain to be around looking
for you within the month, and not with an opera
glass, either.

From Virginia's airy situation one could look over
a vast, far-reaching panorama of mountain ranges
and deserts; and whether the day was bright or
overcast, whether the sun was rising or setting, or
flaming in the zenith, or whether night and the
moon held sway, the spectacle was always impressive
and beautiful. Over your head Mount Davidson lifted
its gray dome, and before and below you a rugged
canyon clove the battlemented hills, making a somber gateway through which a soft-tinted desert was
glimpsed, with the silver thread of a river winding
through it, bordered with trees which many miles of
distance diminished to a delicate fringe; and still
further away the snowy mountains rose up and
stretched their long barrier to the filmy horizon —
far enough beyond a lake that burned in the desert
like a fallen sun, though that, itself, lay fifty miles
removed. Look from your window where you would,
there was fascination in the picture. At rare intervals — but very rare — there were clouds in our skies,
and then the setting sun would gild and flush and
glorify this mighty expanse of scenery with a bewild-
ering pomp of color that held the eye like a spell
and moved the spirit like music.

CHAPTER III.

MY salary was increased to forty dollars a week. But I seldom drew it. I had plenty of other resources, and what were two broad twenty-dollar gold pieces to a man who had his pockets full of such and a cumbersome abundance of bright half dollars besides? [Paper money has never come into use on the Pacific coast.] Reporting was lucrative, and every man in the town was lavish with his money and his "feet." The city and all the great mountain side were riddled with mining shafts. There were more mines than miners. True, not ten of these mines were yielding rock worth hauling to a mill, but everybody said, "Wait till the shaft gets down where the ledge comes in solid, and then you will see!" So nobody was discouraged. These were nearly all "wildcat" mines, and wholly worthless, but nobody believed it then. The "Ophir," the "Gould & Curry," the "Mexican," and other great mines on the Comstock lead in Virginia and Gold Hill were turning out huge piles of rich rock every day, and every man believed that his little wildcat claim was as good as any on the "main lead"

3**

and would infallibly be worth a thousand dollars a foot when he "got down where it came in solid." Poor fellow! he was blessedly blind to the fact that he never would see that day. So the thousand wildcat shafts burrowed deeper and deeper into the earth day by day, and all men were beside themselves with hope and happiness. How they labored, prophesied, exulted! Surely nothing like it was ever seen before since the world began. Every one of these wildcat mines — not mines, but holes in the ground over imaginary mines — was incorporated and had handsomely engraved " stock " and the stock was salable, too. It was bought and sold with a feverish avidity in the boards every day. You could go up on the mountain side, scratch around and find a ledge (there was no lack of them), put up a " notice " with a grandiloquent name on it, start a shaft, get your stock printed, and with nothing whatever to prove that your mine was worth a straw, you could put your stock on the market and sell out for hundreds and even thousands of dollars. To make money, and make it fast, was as easy as it was to eat your dinner. Every man owned " feet " in fifty different wildcat mines and considered his fortune made. Think of a city with not one solitary poor man in it! One would suppose that when month after month went by and still not a wildcat mine (by wildcat I mean, in general terms, *any* claim not located on the mother vein, *i. e.*, the " Comstock ") yielded a ton of rock worth crushing, the people would begin to wonder if they were not putting too much faith in their prospective riches; but there was not a thought of such a thing. They burrowed away, bought and sold, and were happy.

New claims were taken up daily, and it was the friendly custom to run straight to the newspaper offices, give the reporter forty or fifty " feet," and get them to go and examine the mine and publish a notice of it. They did not care a fig what you said about the property so you said something. Consequently we generally said a word or two to the effect that the " indications " were good, or that the ledge was " six feet wide," or that the rock " resembled the Comstock " (and so it did — but as a general thing the resemblance was not startling enough to knock you down). If the rock was moderately promising, we followed the custom of the country, used strong adjectives and frothed at the mouth as if a very marvel in silver discoveries had transpired. If the mine was a " developed " one, and had no pay ore to show (and of course it hadn't), we praised the tunnel; said it was one of the most infatuating tunnels in the land; driveled and driveled about the tunnel till we ran entirely out of ecstasies — but never said a word about the rock. We would squander half a column of adulation on a shaft, or a new wire rope, or a dressed pine windlass, or a fascinating force pump, and close with a burst of admiration of the " gentlemanly and efficient superintendent " of the mine — but never utter a whisper

3..

about the rock. And those people were always pleased, always satisfied. Occasionally we patched up and varnished our reputation for discrimination and stern, undeviating accuracy, by giving some old abandoned claim a blast that ought to have made its dry bones rattle — and then somebody would seize it and sell it on the fleeting notoriety thus conferred upon it.

There was *nothing* in the shape of a mining claim that was not salable. We received presents of "feet" every day. If we needed a hundred dollars or so, we sold some; if not, we hoarded it away, satisfied that it would ultimately be worth a thousand dollars a foot. I had a trunk about half full of "stock." When a claim made a stir in the market and went up to a high figure, I searched through my pile to see if I had any of its stock — and generally found it.

The prices rose and fell constantly; but still a fall disturbed us little, because a thousand dollars a foot was our figure, and so we were content to let it fluctuate as much as it pleased till it reached it. My pile of stock was not all given to me by people who wished their claims "noticed." At least half of it was given me by persons who had no thought of such a thing, and looked for nothing more than a simple verbal "thank you"; and you were not even obliged by law to furnish that. If you are coming up the street with a couple of baskets of apples in your hands, and you meet a friend, you naturally invite him to take a few. That describes the condition of things in Virginia in the " flush times." Every man had his pockets full of stock, and it was the actual *custom* of the country to part with small quantities of it to friends without the asking. Very often it was a good idea to close the transaction instantly, when a man offered a stock present to a friend, for the offer was only good and binding at that moment, and if the price went to a high figure shortly afterward the procrastination was a thing to be regretted. Mr. Stewart (Senator, now, from Nevada) one day told me he would give me twenty feet of " Justis " stock if I would walk over to his office. It was worth five or ten dollars a foot. I asked him to make the offer good for next day, as I was just going to dinner. He said he would not be in town; so I risked it and took my dinner instead of the stock. Within the week the price went up to seventy dollars and afterward to a hundred and fifty, but nothing could make that man yield. I suppose he sold that stock of mine and placed the guilty proceeds in his own pocket. I met three friends one afternoon, who said they had been buying " Overman " stock at auction at eight dollars a foot. One said if I would said he would do the same. But I was going after an inquest and could not stop   A few weeks afterward they sold all their " Overman " at six hundred dollars a foot and generously came around to tell me

C.**

about it — and also to urge me to accept of the next forty-five feet of it that people tried to force on me. These are actual facts, and I could make the list a long one and still confine myself strictly to the truth. Many a time friends gave us as much as twenty-five feet of stock that was selling at twenty-five dollars a foot, and they thought no more of it than they would of offering a guest a cigar. These were " flush times " indeed! I thought they were going to last always, but somehow I never was much of a prophet.

To show what a wild spirit possessed the mining brain of the community, I will remark that " claims " were actually " located " in excavations for cellars, where the pick had exposed what seemed to be quartz veins — and not cellars in the suburbs, either, but in the very heart of the city; and forthwith stock would be issued and thrown on the market. It was small matter who the cellar belonged to — the " ledge " belonged to the finder, and unless the United States government interfered (inasmuch as the government holds the primary right to mines of the noble metals in Nevada — or at least did then), it was considered to be his privilege to work it. Imagine a stranger staking out a mining claim among the costly shrubbery in your front yard and calmly proceeding to lay waste the ground with pick and shovel and blasting powder! It has been often done in California. In the middle of one of the principal business streets of Virginia, a man "located" a mining claim and began a shaft on it. He gave me a hundred feet of the stock and I sold it for a fine suit of clothes because I was afraid somebody would fall down the shaft and sue for damages. I owned in another claim that was located in the middle of another street; and to show how absurd people can be, that "East India" stock (as it was called) sold briskly although there was an ancient tunnel running directly under the claim and any man could go into it and see that it did not cut a quartz ledge or anything that remotely resembled one.

One plan of acquiring sudden wealth was to "salt" a wildcat claim and sell out while the excitement was up. The process was simple. The schemer located a worthless ledge, sunk a shaft on it, bought a wagon load of rich "Comstock" ore, dumped a portion of it into the shaft and piled the rest by its side, above ground. Then he showed the property to a simpleton and sold it to him at a high figure. Of course the wagon load of rich ore was all that the victim ever got out of his purchase. A most remarkable case of "salting" was that of the "North Ophir." It was claimed that this vein was a remote "extension" of the original "Ophir," a valuable mine on the "Comstock." For a few days everybody was talking about the rich developments in the North Ophir. It was said that it yielded perfectly pure silver in small, solid lumps. I went to the place with the owners, and found a shaft six or eight feet deep, in the bottom of which was a badly shattered vein of dull, yellowish, unpromising rock. One would as soon expect to find silver in a grindstone. We got out a pan of the rubbish and washed it in a puddle, and sure enough, among the sediment we found half a dozen black, bullet-looking pellets of unimpeachable "native" silver. Nobody had ever heard of such a thing before; science could not account for such a queer novelty. The stock rose to sixty-five dollars a foot, and at this figure the world-renowned tragedian, McKean Buchanan, bought a commanding interest and prepared to quit the stage once more — he was always doing that. And then it transpired that the mine had been "salted"— and not in any hackneyed way, either, but in a singularly bold, barefaced and peculiarly original and outrageous fashion. On one of the lumps of "native" silver was discovered the minted legend, "TED STATES OF," and then it was plainly apparent that the mine had been "salted" with melted half dollars! The lumps thus obtained had been blackened till they resembled native silver, and were then mixed with the shattered rock in the bottom of the shaft. It is literally true. Of course the price of the stock at once fell to nothing, and the tragedian was ruined. But for this calamity we might have lost McKean Buchanan from the stage.

CHAPTER IV.

THE "flush times" held bravely on. Something over two years before, Mr. Goodman and another journeyman printer had borrowed forty dollars and set out from San Francisco to try their fortunes in the new city of Virginia. They found the *Territorial Enterprise,* a poverty-stricken weekly journal, gasping for breath and likely to die. They bought it, type, fixtures, good-will, and all, for a thousand dollars, on long time. The editorial sanctum, news-room, press-room, publication office, bed-chamber, parlor, and kitchen were all compressed into one apartment, and it was a small one, too. The editors and printers slept on the floor, a Chinaman did their cooking, and the "imposing-stone" was the general dinner table. But now things were changed. The paper was a great daily, printed by steam; there were five editors and twenty-three compositors; the subscription price was sixteen dollars a year; the advertising rates were exorbitant, and the columns crowded. The paper was clearing from six to ten thousand dollars a month, and the "Enterprise Building" was finished and ready for occupation — a stately fire-proof brick.

Every day from five all the way up to eleven columns of " live " advertisements were left out or
crowded into spasmodic and irregular " supple-
ments."

The " Gould & Curry " company were erecting a
monster hundred-stamp mill at a cost that ultimately
fell little short of a million dollars. Gould & Curry
stock paid heavy dividends — a rare thing, and an
experience confined to the dozen or fifteen claims
located on the " main lead," the " Comstock."
The superintendent of the Gould & Curry lived,
rent free, in a fine house built and furnished by the
company. He drove a fine pair of horses which
were a present from the company, and his salary was
twelve thousand dollars a year. The superintendent
of another of the great mines traveled in grand state,
had a salary of twenty-eight thousand dollars a year,
and in a lawsuit in after days claimed that he was
to have had one per cent. of the gross yield of the
bullion likewise.

Money was wonderfully plenty. The trouble was,
not how to get it, — but how to spend it, how to
lavish it, get rid of it, squander it. And so it was
a happy thing that just at this juncture the news
came over the wires that a great United States Sanitary Commission had been formed and money was
wanted for the relief of the wounded sailors and
soldiers of the Union languishing in the Eastern
hospitals. Right on the heels of it came word that
San Francisco had responded superbly before the

telegram was half a day old. Virginia rose as one
man! A Sanitary Committee was hurriedly organized, and its chairman mounted a vacant cart in C
street and tried to make the clamorous multitude
understand that the rest of the committee were flying
hither and thither and working with all their might
and main, and that if the town would only wait an
hour, an office would be ready, books opened, and
the Commission prepared to receive contributions.
His voice was drowned and his information lost in a
ceaseless roar of cheers, and demands that the
money be received *now* — they swore they would not
wait. The chairman pleaded and argued, but, deaf
to all entreaty, men plowed their way through the
throng and rained checks of gold coin into the cart
and scurried away for more. Hands clutching
money were thrust aloft out of the jam by men
who hoped this eloquent appeal would cleave a road
their strugglings could not open. The very Chinamen and Indians caught the excitement and dashed
their half-dollars into the cart without knowing or
caring what it was all about. Women plunged into
the crowd, trimly attired, fought their way to the
cart with their coin, and emerged again, by and by,
with their apparel in a state of hopeless dilapidation.
It was the wildest mob Virginia had ever seen and
the most determined and ungovernable; and when
at last it abated its fury and dispersed, it had not a
penny in its pocket. To use its own phraseology,
it came there " flush " and went away " busted."

After that, the Commission got itself into sys-
tematic working order, and for weeks the contributions flowed into its treasury in a generous stream.
Individuals and all sorts of organizations levied upon
themselves a regular weekly tax for the sanitary
fund, graduated according to their means, and there
was not another grand universal outburst till the
famous " Sanitary Flour Sack " came our way. Its
history is peculiar and interesting. A former schoolmate of mine, by the name of Reuel Gridley, was
living at the little city of Austin, in the Reese river
country, at this time, and was the Democratic candidate for mayor. He and the Republican candidate
made an agreement that the defeated man should be
publicly presented with a fifty-pound sack of flour
by the successful one, and should carry it home on
his shoulder. Gridley was defeated. The new
mayor gave him the sack of flour, and he shouldered
it and carried it a mile or two, from Lower Austin to
his home in Upper Austin, attended by a band of
music and the whole population. Arrived there, he
said he did not need the flour, and asked what the
people thought he had better do with it. A voice
said:

" Sell it to the highest bidder, for the benefit of
the Sanitary fund."

The suggestion was greeted with a round of applause, and Gridley mounted a dry-goods box and
assumed the rôle of auctioneer. The bids went
higher and higher, as the sympathies of the pioneers

awoke and expanded, till at last the sack was
knocked down to a mill man at two hundred and
fifty dollars, and his check taken. He was asked
where he would have the flour delivered, and he
said:

" Nowhere — sell it again.''

Now the cheers went up royally, and the multitude were fairly in the spirit of the thing. So Gridley stood there and shouted and perspired till the sun
went down; and when the crowd dispersed he had
sold the sack to three hundred different people, and
had taken in eight thousand dollars in gold. And
still the flour sack was in his possession.

The news came to Virginia, and a telegram went
back:

" Fetch along your flour sack !''

Thirty-six hours afterward Gridley arrived, and an
afternoon mass meeting was held in the Opera
House, and the auction began. But the sack had
come sooner than it was expected; the people were
not thoroughly aroused, and the sale dragged. At
nightfall only five thousand dollars had been secured,
and there was a crestfallen feeling in the community.
However, there was no disposition to let the matter
rest here and acknowledge vanquishment at the hands
of the village of Austin. Till late in the night the
principal citizens were at work arranging the mor-
row's campaign, and when they went to bed they
had no fears for the result. At eleven the next
morning a procession of open carriages, attended by

clamorous bands of music and adorned with a moving display of flags, filed along C street and was soon
in danger of blockade by a huzzaing multitude of
citizens.   In the first carriage sat Gridley, with the
flour sack in prominent view, the latter splendid with
bright paint and gilt lettering; also in the same carriage sat the mayor and the recorder.   The other
carriages contained the Common Council, the editors
and reporters, and other people of imposing consequence.   The crowd pressed to the corner of C and
Taylor streets, expecting the sale to begin there, but
they were disappointed, and also unspeakably surprised; for the cavalcade moved on as if Virginia
had ceased to be of importance, and took its way
over the "divide," toward the small town of Gold
Hill.   Telegrams had gone ahead to Gold Hill,
Silver City, and Dayton, and those communities were
at fever heat and ripe for the conflict.   It was a very
hot day, and wonderfully dusty.   At the end of a
short half hour we descended into Gold Hill with
drums beating and colors flying, and enveloped in
imposing clouds of dust.   The whole population—
men, women, and children, Chinamen and Indians,
were massed in the main street, all the flags in town
were at the masthead, and the blare of the bands
was drowned in cheers.   Gridley stood up and asked
who would make the first bid for the National Sanitary Flour Sack.   Gen. W. said:

"The Yellow Jacket silver mining company offers
a thousand dollars, coin!"

A tempest of applause followed. A telegram carried the news to Virginia, and fifteen minutes afterward that city's population was massed in the streets
devouring the tidings — for it was part of the program that the bulletin boards should do a good work
that day. Every few minutes a new dispatch was
bulletined from Gold Hill, and still the excitement
grew. Telegrams began to return to us from Virginia beseeching Gridley to bring back the flour
sack; but such was not the plan of the campaign.
At the end of an hour Gold Hill's small population
had paid a figure for the flour sack that awoke all the
enthusiasm of Virginia when the grand total was dis-
played upon the bulletin boards. Then the Gridley
cavalcade moved on, a giant refreshed with new
lager beer and plenty of it — for the people brought
it to the carriages without waiting to measure it —
and within three hours more the expedition had carried Silver City and Dayton by storm and was on its
way back covered with glory. Every move had been
telegraphed and bulletined, and as the procession
entered Virginia and filed down C street at half past
eight in the evening the town was abroad in the
thoroughfares, torches were glaring, flags flying,
bands playing, cheer on cheer cleaving the air, and
the city ready to surrender at discretion. The auction began, every bid was greeted with bursts of applause, and at the end of two hours and a half a
population of fifteen thousand souls had paid in coin
for a fifty-pound sack of flour a sum equal to forty

thousand dollars in greenbacks! It was at a rate in the neighborhood of three dollars for each man, woman, and child of the population. The grand total would have been twice as large, but the streets were very narrow, and hundreds who wanted to bid could not get within a block of the stand, and could not make themselves heard. These grew tired of waiting, and many of them went home long before the auction was over. This was the greatest day Virginia ever saw, perhaps.

Gridley sold the sack in Carson City and several California towns; also in San Francisco. Then he took it East and sold it in one or two Atlantic cities, I think. I am not sure of that, but I know that he finally carried it to St. Louis, where a monster sanitary fair was being held, and after selling it there for a large sum and helping on the enthusiasm by displaying the portly silver bricks which Nevada's donation had produced, he had the flour baked up into small cakes and retailed them at high prices.

It was estimated that when the flour sack's mission was ended it had been sold for a grand total of a hundred and fifty thousand dollars in greenbacks! This is probably the only instance on record where common family flour brought three thousand dollars a pound in the public market.

It is due to Mr. Gridley's memory to mention that the expenses of his Sanitary Flour Sack expedition of fifteen thousand miles, going and returning, were paid in large part, if not entirely, out of his own pocket. The time he gave to it was not less than three months. Mr. Gridley was a soldier in the Mexican war and a pioneer Californian. He died at Stockton, California, in December, 1870, greatly regretted.

4••

CHAPTER V.

THERE were nabobs in those days — in the "flush times," I mean. Every rich strike in the mines created one or two. I call to mind several of these. They were careless, easy-going fellows, as a general thing, and the community at large was as much benefited by their riches as they were themselves — possibly more, in some cases.

Two cousins, teamsters, did some hauling for a man and had to take a small segregated portion of a silver mine in lieu of $300 cash. They gave an outsider a third to open the mine, and they went on teaming. But not long. Ten months afterward the mine was out of debt and paying each owner $8,000 to $10,000 a month — say $100,000 a year.

One of the earliest nabobs that Nevada was delivered of wore $6,000 worth of diamonds in his bosom, and swore he was unhappy because he could not spend his money as fast as he made it.

Another Nevada nabob boasted an income that often reached $16,000 a month; and he used to love to tell how he had worked in the very mine that yielded it, for five dollars a day, when he first came to the country.

(48)

The silver and sage-brush State has knowledge of another of these pets of fortune — lifted from actual poverty to affluence almost in a single night — who was able to offer $100,000 for a position of high official distinction, shortly afterward, and did offer it — but failed to get it, his politics not being as sound as his bank account.

Then there was John Smith. He was a good, honest, kind-hearted soul, born and reared in the lower ranks of life, and miraculously ignorant. He drove a team, and owned a small ranch — a ranch that paid him a comfortable living, for although it yielded but little hay, what little it did yield was worth from $250 to $300 in gold per ton in the market. Presently Smith traded a few acres of the ranch for a small undeveloped silver mine in Gold Hill. He opened the mine and built a little unpretending ten-stamp mill. Eighteen months afterward he retired from the hay business, for his mining income had reached a most comfortable figure. Some people said it was $30,000 a month, and others said it was $60,000. Smith was very rich, at any rate.

And then he went to Europe and traveled. And when he came back he was never tired of telling about the fine hogs he had seen in England, and the gorgeous sheep he had seen in Spain, and the fine cattle he had noticed in the vicinity of Rome. He was full of the wonders of the old world, and advised everybody to travel. He said a man never imagined

4 **

what surprising things there were in the world till he had traveled.

One day, on board ship, the passengers made up a pool of $500, which was to be the property of the man who should come nearest to guessing the run of the vessel for the next twenty-four hours. Next day, toward noon, the figures were all in the purser's hands in sealed envelopes. Smith was serene and happy, for he had been bribing the engineer. But another party won the prize! Smith said:

" Here, that won't do! He guessed two miles wider of the mark than I did."

The purser said, " Mr. Smith, you missed it further than any man on board. We traveled two hundred and eight miles yesterday."

" Well, sir," said Smith, " that's just where I've got you, for I guessed two hundred and nine. If you'll look at my figgers again you'll find a 2 and two 0's, which stands for 200, don't it? — and after 'em you'll find a 9 (2009), which stands for two hundred and nine. I reckon I'll take that money, if you please."

The Gould & Curry claim comprised twelve hundred feet, and it all belonged originally to the two men whose names it bears. Mr. Curry owned two-thirds of it — and he said that he sold it out for twenty-five hundred dollars in cash, and an old plug horse that ate up his market value in hay and barley in seventeen days by the watch. And he said that

Gould sold out for a pair of second-hand government blankets and a bottle of whisky that killed nine
men in three hours, and that an unoffending stranger
that smelt the cork was disabled for life. Four
years afterward the mine thus disposed of was worth
in the San Francisco market seven millions six hundred thousand dollars in gold coin.

In the early days a poverty-stricken Mexican who
lived in a canyon directly back of Virginia City, had
a stream of water as large as a man's wrist trickling
from the hillside on his premises. The Ophir Company segregated a hundred feet of their mine and
traded it to him for the stream of water. The
hundred feet proved to be the richest part of the
entire mine; four years after the swap, its market
value (including its mill) was $1,500,000.

An individual who owned twenty feet in the
Ophir mine before its great riches were revealed to
men, traded it for a horse, and a very sorry-looking
brute he was, too. A year or so afterward, when
Ophir stock went up to $3,000 a foot, this man,
who had not a cent, used to say he was the most
startling example of magnificence and misery the
world had ever seen — because he was able to ride
a sixty-thousand-dollar horse — yet could not scrape
up cash enough to buy a saddle, and was obliged to
borrow one or ride bareback. He said if fortune
were to give him another sixty-thousand-dollar
horse it would ruin him.

A youth of nineteen, who was a telegraph

D**

operator in Virginia on a salary of a hundred dollars
a month, and who, when he could not make out
German names in the list of San Francisco steamer
arrivals, used to ingeniously select and supply sub-
stitutes for them out of an old Berlin city directory,
made himself rich by watching the mining telegrams
that passed through his hands and buying and selling stocks accordingly, through a friend in San
Francisco. Once when a private dispatch was sent
from Virginia announcing a rich strike in a prominent mine and advising that the matter be kept
secret till a large amount of the stock could be
secured, he bought forty "feet" of the stock at
twenty dollars a foot, and afterward sold half of it
at eight hundred dollars a foot and the rest at
double that figure. Within three months he was
worth $150,000, and had resigned his telegraphic
position.

Another telegraph operator who had been discharged by the company for divulging the secrets of
the office, agreed with a moneyed man in San
Francisco to furnish him the result of a great Virginia mining lawsuit within an hour after its private
reception by the parties to it in San Francisco. For
this he was to have a large percentage of the profits
on purchases and sales made on it by his fellow-
conspirator. So he went, disguised as a teamster,
to a little wayside telegraph office in the mountains,
got acquainted with the operator, and sat in the
office day after day, smoking his pipe, complaining

that his team was fagged out and unable to travel — and meantime listening to the dispatches as they passed clicking through the machine from Virginia. Finally the private dispatch announcing the result of the lawsuit sped over the wires, and as soon as he heard it he telegraphed his friend in San Francisco :

" Am tired waiting. Shall sell the team and go home."

It was the signal agreed upon. The word " waiting " left out, would have signified that the suit had gone the other way. The mock teamster's friend picked up a deal of the mining stock, at low figures, before the news became public, and a fortune was the result.

For a long time after one of the great Virginia mines had been incorporated, about fifty feet of the original location were still in the hands of a man who had never signed the incorporation papers. The stock became very valuable, and every effort was made to find this man, but he had disappeared. Once it was heard that he was in New York, and one or two speculators went East but failed to find him. Once the news came that he was in the Bermudas, and straightway a speculator or two hurried east and sailed for Bermuda — but he was not there. Finally he was heard of in Mexico, and a friend of his, a barkeeper on a salary, scraped together a little money and sought him out, bought his " feet " for a hundred dollars, returned and sold the property for $75,000.

But why go on? The traditions of Silverland are filled with instances like these, and I would never get through enumerating them were I to attempt to do it. I only desired to give the reader an idea of a peculiarity of the " flush times " which I could not present so strikingly in any other way, and which some mention of was necessary to a realizing comprehension of the time and the country.

I was personally acquainted with the majority of the nabobs I have referred to, and so, for old acquaintance's sake, I have shifted their occupations and experiences around in such a way as to keep the Pacific public from recognizing these once notorious men. No longer notorious, for the majority of them have drifted back into poverty and obscurity again.

In Nevada there used to be current the story of an adventure of two of her nabobs, which may or may not have occurred. I give it for what it is worth:

Col. Jim had seen somewhat of the world, and knew more or less of its ways; but Col. Jack was from the back settlements of the States, had led a life of arduous toil, and had never seen a city. These two, blessed with sudden wealth, projected a visit to New York,— Col. Jack to see the sights, and Col. Jim to guard his unsophistication from misfortune. They reached San Francisco in the night, and sailed in the morning. Arrived in New York, Col. Jack said:

" I've heard tell of carriages all my life, and now I mean to have a ride in one; I don't care what it costs. Come along.''

They stepped out on the sidewalk, and Col. Jim called a stylish barouche. But Col. Jack said:

"*No*, sir! None of your cheap-John turnouts for me. I'm here to have a good time, and money ain't any object. I mean to have the nobbiest rig that's going. Now here comes the very trick. Stop that yaller one with the pictures on it — don't you fret — I'll stand all the expenses myself.''

So Col. Jim stopped an empty omnibus, and they got in. Said Col. Jack:

" Ain't it gay, though? Oh, no, I reckon not! Cushions, and windows, and pictures, till you can't rest. What would the boys say if they could see us cutting a swell like this in New York? By George, I wish they *could* see us.''

Then he put his head out of the window, and shouted to the driver:

" Say, Johnny, this suits *me!* — suits yours truly, you bet, you! I want this shebang all day. I'm *on* it, old man! Let 'em out! Make 'em go! We'll make it all right with *you*, sonny!''

The driver passed his hand through the strap-hole, and tapped for his fare — it was before the gongs came into common use. Col. Jack took the hand, and shook it cordially. He said:

" You twig me, old pard! All right between gents. Smell of *that*, and see how you like it!''

And he put a twenty-dollar gold piece in the driver's hand. After a moment the driver said he could not make change.

"Bother the change! Ride it out. Put it in your pocket."

Then to Col. Jim, with a sounding slap on his thigh:

"*Ain't* it style, though? Hanged if I don't hire this thing every day for a week."

The omnibus stopped, and a young lady got in. Col. Jack stared a moment, then nudged Col. Jim with his elbow:

"Don't say a word," he whispered. "Let her ride, if she wants to. Gracious, there's room enough."

The young lady got out her portemonnaie, and handed her fare to Col. Jack.

"What's this for?" said he.

"Give it to the driver, please."

"Take back your money, madam. We can't allow it. You're welcome to ride here as long as you please, but this shebang's chartered, and we can't let you pay a cent."

The girl shrunk into a corner, bewildered. An old lady with a basket climbed in, and proffered her fare.

"Excuse me," said Col. Jack. "You're perfectly welcome here, madam, but we can't allow you to pay. Set right down there, mum, and don't you be the least uneasy. Make yourself just as free as if you was in your own turnout."

Within two minutes, three gentlemen, two fat women, and a couple of children, entered.

"Come right along, friends," said Col. Jack; "don't mind *us*. This is a free blowout." Then he whispered to Col. Jim, "New York ain't no sociable place, I don't reckon — it ain't no *name* for it!"

He resisted every effort to pass fares to the driver, and made everybody cordially welcome. The situation dawned on the people, and they pocketed their money, and delivered themselves up to covert enjoyment of the episode. Half a dozen more passengers entered.

"Oh, there's *plenty* of room," said Col. Jack. "Walk right in, and make yourselves at home. A blowout ain't worth anything *as* a blowout, unless a body has company." Then in a whisper to Col. Jim: "But *ain't* these New Yorkers friendly? And ain't they cool about it, too? Icebergs ain't anywhere. I reckon they'd tackle a hearse, if it was going their way."

More passengers got in; more yet, and still more. Both seats were filled, and a file of men were standing up, holding on to the cleats overhead. Parties with baskets and bundles were climbing up on the roof. Half-suppressed laughter rippled up from all sides.

"Well, for clean, cool, out-and-out cheek, if this don't bang anything that ever I saw, I'm an Injun!" whispered Col. Jack.

A Chinaman crowded his way in.

"I weaken!" said Col. Jack. "Hold on, driver! Keep your seats, ladies and gents. Just make yourselves free — everything's paid for. Driver, rustle these folks around as long as they're a mind to go — friends of ours, you know. Take them everywheres — and if you want more money, come to the St. Nicholas, and we'll make it all right. Pleasant journey to you, ladies and gents — go it just as long as you please — it shan't cost you a cent!"

The two comrades got out, and Col. Jack said:

"Jimmy, it's the sociablest place *I* ever saw. The Chinaman waltzed in as comfortable as anybody. If we'd staid awhile, I reckon we'd had some niggers. B' George, we'll have to barricade our doors to-night, or some of these ducks will be trying to sleep with us."

CHAPTER VI.

SOMEBODY has said that in order to know a community, one must observe the style of its funerals and know what manner of men they bury with most ceremony. I cannot say which class we buried with most éclat in our "flush times," the distinguished public benefactor or the distinguished rough — possibly the two chief grades or grand divisions of society honored their illustrious dead about equally; and hence, no doubt, the philosopher I have quoted from would have needed to see two representative funerals in Virginia before forming his estimate of the people.

There was a grand time over Buck Fanshaw when he died. He was a representative citizen. He had "killed his man" — not in his own quarrel, it is true, but in defence of a stranger unfairly beset by numbers. He had kept a sumptuous saloon. He had been the proprietor of a dashing helpmeet whom he could have discarded without the formality of a divorce. He had held a high position in the fire department and been a very Warwick in politics. When he died there was great lamentation through-

out the town, but especially in the vast bottom-
stratum of society.

On the inquest it was shown that Buck Fanshaw,
in the delirium of a wasting typhoid fever, had
taken arsenic, shot himself through the body, cut
his throat, and jumped out of a four-story window
and broken his neck — and after due deliberation,
the jury, sad and tearful, but with intelligence un-
blinded by its sorrow, brought in a verdict of death
"by the visitation of God." What could the world
do without juries?

Prodigious preparations were made for the
funeral. All the vehicles in town were hired, all
the saloons put in mourning, all the municipal and
fire-company flags hung at half-mast, and all the
firemen ordered to muster in uniform and bring
their machines duly draped in black. Now — let
us remark in parenthesis — as all the peoples of the
earth had representative adventurers in the Silverland, and as each adventurer had brought the slang
of his nation or his locality with him, the combination made the slang of Nevada the richest and the
most infinitely varied and copious that had ever
existed anywhere in the world, perhaps, except in
the mines of California in the "early days." Slang
was the language of Nevada. It was hard to preach
a sermon without it, and be understood. Such
phrases as "You bet!" "Oh, no, I reckon not!"
"No Irish need apply," and a hundred others, became so common as to fall from the lips of a

speaker unconsciously — and very often when they
did not touch the subject under discussion and consequently failed to mean anything.

After Buck Fanshaw's inquest, a meeting of the
short-haired brotherhood was held, for nothing can
be done on the Pacific coast without a public meeting and an expression of sentiment. Regretful
resolutions were passed and various committees appointed; among others, a committee of one was
deputed to call on the minister, a fragile, gentle,
spirituel new fledgling from an Eastern theological
seminary, and as yet unacquainted with the ways of
the mines. The committeeman, "Scotty" Briggs,
made his visit; and in after days it was worth something to hear the minister tell about it. Scotty was
a stalwart rough, whose customary suit, when on
weighty official business, like committee work, was
a fire helmet, flaming red flannel shirt, patent leather
belt with spanner and revolver attached, coat hung
over arm, and pants stuffed into boot tops. He
formed something of a contrast to the pale theological student. It is fair to say of Scotty, however,
in passing, that he had a warm heart, and a strong
love for his friends, and never entered into a quarrel
when he could reasonably keep out of it. Indeed,
it was commonly said that whenever one of Scotty's
fights was investigated, it always turned out that it
had originally been no affair of his, but that out of
native goodheartedness he had dropped in of his
own accord to help the man who was getting the

worst of it.   He and Buck Fanshaw were bosom
friends, for years, and had often taken adventurous
" potluck " together.   On one occasion, they had
thrown off their coats and taken the weaker side in
a fight among strangers, and after gaining a hard-
earned victory, turned and found that the men they
were helping had deserted early, and not only that,
but had stolen their coats and made off with them!
But to return to Scotty's visit to the minister.   He
was on a sorrowful mission, now, and his face was
the picture of woe.   Being admitted to the presence
he sat down before the clergyman, placed his fire-
hat on an unfinished manuscript sermon under the
minister's nose, took from it a red silk handkerchief,
wiped his brow and heaved a sigh of dismal impres-
siveness, explanatory of his business.   He choked,
and even shed tears; but with an effort he mastered
his voice and said in lugubrious tones:

" Are you the duck that runs the gospel-mill next
door?"

"Am I the — pardon me, I believe I do not
understand?"

With another sigh and a half-sob, Scotty rejoined:

" Why you see we are in a bit of trouble, and the
boys thought maybe you would give us a lift, if
we'd tackle you — that is, if I've got the rights of
it and you are the head clerk of the doxology-works
next door."

" I am the shepherd in charge of the flock whose
fold is next door."

"The which?"

"The spiritual adviser of the little company of believers whose sanctuary adjoins these premises."

Scotty scratched his head, reflected a moment, and then said:

"You ruther hold over me, pard. I reckon I can't call that hand. Ante and pass the buck."

"How? I beg pardon. What did I understand you to say?"

"Well, you've ruther got the bulge on me. Or maybe we've both got the bulge, somehow. You don't smoke me and I don't smoke you. You see, one of the boys has passed in his checks, and we want to give him a good send-off, and so the thing I'm on now is to roust out somebody to jerk a little chin-music for us and waltz him through handsome."

"My friend, I seem to grow more and more bewildered. Your observations are wholly incomprehensible to me. Cannot you simplify them in some way? At first I thought perhaps I understood you, but I grope now. Would it not expedite matters if you restricted yourself to categorical statements of fact unencumbered with obstructing accumulations of metaphor and allegory?"

Another pause, and more reflection. Then, said Scotty:

"I'll have to pass, I judge."

"How?"

"You've raised me out, pard."

" I still fail to catch your meaning."

" Why, that last lead of yourn is too many for
me — that's the idea. I can't neither trump nor
follow suit."

The clergyman sank back in his chair perplexed.
Scotty leaned his head on his hand and gave himself
up to thought. Presently his face came up, sorrowful but confident.

" I've got it now, so's you can savvy," he said.
" What we want is a gospel-sharp. See?"

" A what?"

" Gospel-sharp. Parson."

" Oh! Why did you not say so before? I am a
clergyman — a parson."

" Now you talk! You see my blind and straddle
it like a man. Put it there!"— extending a brawny
paw, which closed over the minister's small hand
and gave it a shake indicative of fraternal sympathy
and fervent gratification.

" Now we're all right, pard. Let's start fresh.
Don't you mind my snuffling a little — becuz we're
in a power of trouble. You see, one of the boys
has gone up the flume —"

" Gone where?"

" Up the flume — throwed up the sponge, you
understand."

" Thrown up the sponge?"

" Yes — kicked the bucket —"

" Ah — has departed to that mysterious country
from whose bourne no traveler returns."

"Return! I reckon not. Why, pard, he's *dead!*"

"Yes, I understand."

"Oh, you do? Well I thought maybe you might be getting tangled some more. Yes, you see he's dead again —"

"*Again!* Why, has he ever been dead before?"

"Dead before? No! Do you reckon a man has got as many lives as a cat? But you bet you he's awful dead now, poor old boy, and I wish I'd never seen this day. I don't want no better friend than Buck Fanshaw. I knowed him by the back; and when I know a man and like him, I freeze to him — you hear *me*. Take him all round, pard, there never was a bullier man in the mines. No man ever knowed Buck Fanshaw to go back on a friend. But it's all up, you know, it's all up. It ain't no use. They've scooped him."

"Scooped him?"

"Yes — death has. Well, well, well, we've got to give him up. Yes, indeed. It's a kind of a hard world, after all, *ain't* it? But pard, he was a rustler! You ought to seen him get started once. He was a bully boy with a glass eye! Just spit in his face and give him room according to his strength, and it was just beautiful to see him peel and go in. He was the worst son of a thief that ever drawed breath. Pard, he was *on* it! He was on it bigger than an Injun!"

"On it? On what?"

5**

"On the shoot. On the shoulder. On the fight, you understand. *He* didn't give a continental for *any*body. *Beg* your pardon, friend, for coming so near saying a cuss-word — but you see I'm on an awful strain, in this palaver, on account of having to cramp down and draw everything so mild. But we've got to give him up. There ain't any getting around that, I don't reckon. Now if we can get you to help plant him —"

"Preach the funeral discourse? Assist at the obsequies?"

"Obs'quies is good. Yes. That's it — that's our little game. We are going to get the thing up regardless, you know. He was always nifty himself, and so you bet you his funeral ain't going to be no slouch — solid silver door-plate on his coffin, six plumes on the hearse, and a nigger on the box in a biled shirt and a plug hat — how's that for high? And we'll take care of *you*, pard. We'll fix you all right. There'll be a kerridge for you; and whatever you want, you just 'scape out and we'll 'tend to it. We've got a shebang fixed up for you to stand behind, in No. 1's house, and don't you be afraid. Just go in and toot your horn, if you don't sell a clam. Put Buck through as bully as you can, pard, for anybody that knowed him will tell you that he was one of the whitest men that was ever in the mines. You can't draw it too strong. He never could stand it to see things going wrong. He's done more to make this town quiet and

peaceable than any man in it. I've seen him lick
four Greasers in eleven minutes, myself. If a thing
wanted regulating, *he* warn't a man to go browsing
around after somebody to do it, but he would prance
in and regulate it himself. He warn't a Catholic.
Scasely. He was down on 'em. His word was,
'No Irish need apply!' But it didn't make no
difference about that when it came down to what a
man's rights was — and so, when some roughs
jumped the Catholic boneyard and started in to
stake out town-lots in it he *went* for 'em! And he
cleaned 'em, too! I was there, pard, and I seen it
myself."

"That was very well indeed — at least the impulse was — whether the act was strictly defensible
or not. Had deceased any religious convictions?
That is to say, did he feel a dependence upon, or
acknowledge allegiance to a higher power?"

More reflection.

"I reckon you've stumped me again, pard.
Could you say it over once more, and say it slow?"

"Well, to simplify it somewhat, was he, or rather
had he ever been connected with any organization
sequestered from secular concerns and devoted to
self-sacrifice in the interests of morality?"

"All down but nine — set 'em up on the other
alley, pard."

"What did I understand you to say?"

"Why, you're most too many for me, you know.
When you get in with your left I hunt grass every

E **

time. Every time you draw, you fill; but I don't seem to have any luck. Let's have a new deal."

" How? Begin again?"

" That's it."

" Very well. Was he a good man, and —"

" There — I see that; don't put up another chip till I look at my hand. A good man, says you? Pard, it ain't no name for it. He was the best man that ever — pard, you would have doted on that man. He could lam any galoot of his inches in America. It was him that put down the riot last election before it got a start; and everybody said he was the only man that could have done it. He waltzed in with a spanner in one hand and a trumpet in the other, and sent fourteen men home on a shutter in less than three minutes. He had that riot all broke up and prevented nice before anybody ever got a chance to strike a blow. He was always for peace, and he would *have* peace — he could not stand disturbances. Pard, he was a great loss to this town. It would please the boys if you could chip in something like that and do him justice. Here once when the Micks got to throwing stones through the Methodis' Sunday-school windows, Buck Fanshaw, all of his own notion, shut up his saloon and took a couple of six-shooters and mounted guard over the Sunday-school. Says he, ' No Irish need apply!' And they didn't. He was the bulliest man in the mountains, pard! He could run faster, jump higher, hit harder, and hold

more tanglefoot whisky without spilling it than any
man in seventeen counties. Put that in, pard —
it'll please the boys more than anything you could
say. And you can say, pard, that he never shook
his mother."

"Never shook his mother?"

"That's it — any of the boys will tell you so."

"Well, but why *should* he shake her?"

"That's what *I* say — but some people does."

"Not people of any repute?"

"Well, some that averages pretty so-so."

"In my opinion the man that would offer personal violence to his own mother, ought to —"

"Cheese it, pard; you've banked your ball clean
outside the string. What I was a drivin' at, was,
that he never *throwed off* on his mother — don't you
see? No indeedy. He give her a house to live in,
and town lots, and plenty of money; and he looked
after her and took care of her all the time; and
when she was down with the smallpox I'm d—d if
he didn't set up nights and nuss her himself! *Beg*
your pardon for saying it, but it hopped out too
quick for yours truly. You've treated me like a
gentleman, pard, and I ain't the man to hurt your
feelings intentional. I think you're white. I think
you're a square man, pard. I like you, and I'll
lick any man that don't. I'll lick him till he can't
tell himself from a last year's corpse! Put it
there!" [Another fraternal hand-shake — and exit.]

The obsequies were all that "the boys" could

desire. Such a marvel of funeral pomp had never
been seen in Virginia. The plumed hearse, the
dirge-breathing brass bands, the closed marts of
business, the flags drooping at half mast, the long,
plodding procession of uniformed secret societies,
military battalions and fire companies, draped en-
gines, carriages of officials, and citizens in vehicles
and on foot, attracted multitudes of spectators to
the sidewalks, roofs, and windows; and for years
afterward, the degree of grandeur attained by any
civic display in Virginia was determined by comparison with Buck Fanshaw's funeral.

Scotty Briggs, as a pall-bearer and a mourner,
occupied a prominent place at the funeral, and when
the sermon was finished and the last sentence of the
prayer for the dead man's soul ascended, he responded, in a low voice, but with feeling:

"Amen. No Irish need apply."

As the bulk of the response was without apparent
relevancy, it was probably nothing more than a
humble tribute to the memory of the friend that was
gone; for, as Scotty had once said, it was "his
word."

Scotty Briggs, in after days, achieved the distinction of becoming the only convert to religion that
was ever gathered from the Virginia roughs; and it
transpired that the man who had it in him to espouse
the quarrel of the weak out of inborn nobility of
spirit was no mean timber whereof to construct a
Christian. The making him one did not warp his

generosity or diminish his courage; on the contrary
it gave intelligent direction to the one and a broader
field to the other. If his Sunday-school class progressed faster than the other classes, was it matter
for wonder? I think not. He talked to his pioneer
small-fry in a language they understood! It was
my large privilege, a month before he died, to hear
him tell the beautiful story of Joseph and his
brethren to his class " without looking at the book."
I leave it to the reader to fancy what it was like, as
it fell, riddled with slang, from the lips of that grave,
earnest teacher, and was listened to by his little
learners with a consuming interest that showed that
they were as unconscious as he was that any violence was being done to the sacred proprieties!

CHAPTER VII.

THE first twenty-six graves in the Virginia cemetery were occupied by *murdered* men. So
everybody said, so everybody believed, and so they
will always say and believe. The reason why there
was so much slaughtering done, was, that in a new
mining district the rough element predominates, and
a person is not respected until he has "killed his
man." That was the very expression used.

If an unknown individual arrived, they did not
inquire if he was capable, honest, industrious, but—
had he killed his man? If he had not, he gravitated
to his natural and proper position, that of a man of
small consequence; if he had, the cordiality of his
reception was graduated according to the number of
his dead. It was tedious work struggling up to a
position of influence with bloodless hands; but
when a man came with the blood of half-a-dozen
men on his soul, his worth was recognized at once
and his acquaintance sought.

In Nevada, for a time, the lawyer, the editor, the
banker, the chief desperado, the chief gambler, and
the saloon-keeper, occupied the same level in
society, and it was the highest. The cheapest and

easiest way to become an influential man and be
looked up to by the community at large, was to
stand behind a bar, wear a cluster-diamond pin, and
sell whisky. I am not sure but that the saloon-
keeper held a shade higher rank than any other
member of society. His opinion had weight. It
was his privilege to say how the elections should go.
No great movement could succeed without the
countenance and direction of the saloon-keepers. It
was a high favor when the chief saloon-keeper consented to serve in the legislature or the board of
aldermen. Youthful ambition hardly aspired so
much to the honors of the law, or the army and
navy as to the dignity of proprietorship in a saloon.
To be a saloon-keeper and kill a man was to be
illustrious. Hence the reader will not be surprised
to learn that more than one man was killed in
Nevada under hardly the pretext of provocation, so
impatient was the slayer to achieve reputation and
throw off the galling sense of being held in indifferent repute by his associates. I knew two youths
who tried to " kill their men " for no other reason
— and got killed themselves for their pains.
" There goes the man that killed Bill Adams " was
higher praise and a sweeter sound in the ears of this
sort of people than any other speech that admiring
lips could utter.

The men who murdered Virginia's original
twenty-six cemetery-occupants were never punished.
Why? Because Alfred the Great, when he invented

trial by jury, and knew that he had admirably framed it to secure justice in his age of the world, was not aware that in the nineteenth century the condition of things would be so entirely changed that unless he rose from the grave and altered the jury plan to meet the emergency, it would prove the most ingenious and infallible agency for *defeating* justice that human wisdom could contrive. For how could he imagine that we simpletons would go on using his jury plan after circumstances had stripped it of its usefulness, any more than he could imagine that we would go on using his candle-clock after we had invented chronometers? In his day news could not travel fast, and hence he could easily find a jury of honest, intelligent men who had not heard of the case they were called to try — but in our day of telegraphs and newspapers his plan compels us to swear in juries composed of fools and rascals, because the system rigidly excludes honest men and men of brains.

I remember one of those sorrowful farces, in Virginia, which we call a jury trial. A noted desperado killed Mr. B., a good citizen, in the most wanton and cold-blooded way. Of course the papers were full of it, and all men capable of reading read about it. And of course all men not deaf and dumb and idiotic talked about it. A jury list was made out, and Mr. B. L., a prominent banker and a valued citizen, was questioned precisely as he would have been questioned in any court in America:

" Have you heard of this homicide?"

" Yes."

" Have you held conversations upon the subject?"

" Yes."

" Have you formed or expressed opinions about it?"

" Yes."

" Have you read the newspaper accounts of it?"

" Yes."

" We do not want you."

A minister, intelligent, esteemed, and greatly respected; a merchant of high character and known probity; a mining superintendent of intelligence and unblemished reputation; a quartz mill owner of excellent standing, were all questioned in the same way, and all set aside. Each said the public talk and the newspaper reports had not so biased his mind but that sworn testimony would overthrow his previously-formed opinions and enable him to render a verdict without prejudice and in accordance with the facts. But of course such men could not be trusted with the case. Ignoramuses alone could mete out unsullied justice.

When the peremptory challenges were all exhausted, a jury of twelve men was impaneled — a jury who swore they had neither heard, read, talked about, nor expressed an opinion concerning a murder which the very cattle in the corrals, the Indians in the sage-brush, and the stones in the streets were

cognizant of! It was a jury composed of two desperadoes, two low beer-house politicians, three bar-
keepers, two ranchmen who could not read, and
three dull, stupid, human donkeys! It actually
came out afterward, that one of these latter thought
that incest and arson were the same thing.

The verdict rendered by this jury was, Not Guilty.
What else could one expect?

The jury system puts a ban upon intelligence and
honesty, and a premium upon ignorance, stupidity,
and perjury. It is a shame that we must continue
to use a worthless system because it *was* good a
thousand years ago. In this age, when a gentleman
of high social standing, intelligence, and probity,
swears that testimony given under solemn oath will
outweigh, with him, street talk and newspaper reports based upon mere hearsay, he is worth a hundred jurymen who will swear to their own ignorance
and stupidity, and justice would be far safer in his
hands than in theirs. Why could not the jury law
be so altered as to give men of brains and honesty
an *equal chance* with fools and miscreants? Is it
right to show the present favoritism to one class of
men and inflict a disability on another, in a land
whose boast is that all its citizens are free and
equal? I am a candidate for the legislature. I
desire to tamper with the jury law. I wish to so
alter it as to put a premium on intelligence and
character, and close the jury box against idiots,
blacklegs, and people who do not read newspapers.

But no doubt I shall be defeated — every effort I make to save the country "misses fire."

My idea, when I began this chapter, was to say something about desperadoism in the "flush times" of Nevada. To attempt a portrayal of that era and that land, and leave out the blood and carnage, would be like portraying Mormondom and leaving out polygamy. The desperado stalked the streets with a swagger graded according to the number of his homicides, and a nod of recognition from him was sufficient to make a humble admirer happy for the rest of the day. The deference that was paid to a desperado of wide reputation, and who "kept his private graveyard," as the phrase went, was marked, and cheerfully accorded. When he moved along the sidewalk in his excessively long-tailed frock-coat, shiny stump-toed boots, and with dainty little slouch hat tipped over left eye, the small-fry roughs made room for his majesty; when he entered the restaurant, the waiters deserted bankers and merchants to overwhelm him with obsequious service; when he shouldered his way to a bar, the shouldered parties wheeled indignantly, recognized him, and — apologized. They got a look in return that froze their marrow, and by that time a curled and breast-pinned bar-keeper was beaming over the counter, proud of the established acquaintanceship that permitted such a familiar form of speech as:

"How 're ye, Billy, old fel? Glad to see you. What'll you take — the old thing?"

The " old thing " meant his customary drink, of course.

The best-known names in the territory of Nevada were those belonging to these long-tailed heroes of the revolver. Orators, governors, capitalists, and leaders of the legislature enjoyed a degree of fame, but it seemed local and meager when contrasted with the fame of such men as Sam Brown, Jack Williams, Billy Mulligan, Farmer Pease, Sugarfoot Mike, Pock-Marked Jake, El Dorado Johnny, Jack McNabb, Joe McGee, Jack Harris, Six-fingered Pete, etc., etc. There was a long list of them. They were brave, reckless men, and traveled with their lives in their hands. To give them their due, they did their killing principally among themselves, and seldom molested peaceable citizens, for they considered it small credit to add to their trophies so cheap a bauble as the death of a man who was " not on the shoot," as they phrased it. They killed each other on slight provocation, and hoped and expected to be killed themselves — for they held it almost shame to die otherwise than " with their boots on," as they expressed it.

I remember an instance of a desperado's contempt for such small game as a private citizen's life. I was taking a late supper in a restaurant one night, with two reporters and a little printer named — Brown, for instance — any name will do. Presently a stranger with a long-tailed coat on came in, and not noticing Brown's hat, which was lying in a chair,

sat down on it. Little Brown sprang up and became abusive in a moment. The stranger smiled,
smoothed out the hat, and offered it to Brown with
profuse apologies couched in caustic sarcasm, and
begged Brown not to destroy him. Brown threw
off his coat and challenged the man to fight —
abused him, threatened him, impeached his courage,
and urged and even implored him to fight; and in
the meantime the smiling stranger placed himself
under our protection in mock distress. But presently he assumed a serious tone, and said:

" Very well, gentlemen, if we must fight, we must,
I suppose. But don't rush into danger and then
say I gave you no warning. I am more than a
match for all of you when I get started. I will give
you proofs, and then if my friend here still insists,
I will try to accommodate him."

The table we were sitting at was about five feet
long, and unusually cumbersome and heavy. He
asked us to put our hands on the dishes and hold
them in their places a moment — one of them was a
large oval dish with a portly roast on it. Then he
sat down, tilted up one end of the table, set two of
the legs on his knees, took the end of the table between his teeth, took his hands away, and pulled
down with his teeth till the table came up to a level
position, dishes and all! He said he could lift a
keg of nails with his teeth. He picked up a common glass tumbler and bit a semicircle out of it.
Then he opened his bosom and showed us a net-

6**

work of knife and bullet scars; showed us more on his arms and face, and said he believed he had bullets enough in his body to make a pig of lead. He was armed to the teeth. He closed with the remark that he was Mr. —— of Cariboo — a celebrated name whereat we shook in our shoes. I would publish the name, but for the suspicion that he might come and carve me. He finally inquired if Brown still thirsted for blood. Brown turned the thing over in his mind a moment, and then — asked him to supper.

With the permission of the reader, I will group together, in the next chapter, some samples of life in our small mountain village in the old days of desperadoism. I was there at the time. The reader will observe peculiarities in our *official* society; and he will observe also, an instance of how, in new countries, murders breed murders.

CHAPTER VIII.

AN extract or two from the newspapers of the day will furnish a photograph that can need no embellishment:

FATAL SHOOTING AFFRAY.—An affray occurred, last evening, in a billiard saloon on C street, between *Deputy Marshal Jack Williams* and Wm. Brown, which resulted in the immediate death of the latter. There had been some difficulty between the parties for several months.

An inquest was immediately held, and the following testimony adduced:

Officer GEO. BIRDSALL, sworn, says :—I was told Wm. Brown was drunk and was looking for Jack Williams ; so soon as I heard that I started for the parties to prevent a collision ; went into the billiard saloon ; saw Billy Brown running around, saying if anybody had anything against him to show cause ; he was talking in a boisterous manner, and officer Perry took him to the other end of the room to talk to him ; Brown came back to me ; remarked to me that he thought he was as good as anybody, and knew how to take care of himself ; he passed by me and went to the bar ; don't know whether he drank or not ; Williams was at the end of the billiard-table, next to the stairway ; Brown, after going to the bar, came back and said he was as good as any man in the world ; he had then walked out to the end of the first billiard-table from the bar ; I moved closer to them, supposing there would be a fight ; as Brown drew his pistol I caught hold of it ; he had fired one shot at Williams ; don't know the effect of it ; caught hold of him with one hand, and took hold of the pistol and turned it up ; think he fired once after I caught hold of the pistol ; I wrenched the pistol from him ; walked to the end of the billiard-table and told a party that I had Brown's pistol, and to stop shooting ; I think four shots were fired in all ; after walking out, Mr. Foster remarked that Brown was shot dead.

6**   (81)

Oh, there was no excitement about it — he merely "remarked" the small circumstance!

Four months later the following item appeared in the same paper (the *Enterprise*). In this item the name of one of the city officers above referred to (*Deputy Marshal Jack Williams*) occurs again:

ROBBERY AND DESPERATE AFFRAY.—On Tuesday night, a German named Charles Hurtzal, engineer in a mill at Silver City, came to this place, and visited the hurdy-gurdy house on B street. The music, dancing, and Teutonic maidens awakened memories of Faderland until our German friend was carried away with rapture. He evidently had money, and was spending it freely. Late in the evening Jack Williams and Andy Blessington invited him down stairs to take a cup of coffee. Williams proposed a game of cards and went up stairs to procure a deck, but not finding any returned. On the stairway he met the German, and drawing his pistol knocked him down and rifled his pockets of some seventy dollars. Hurtzal dared give no alarm, as he was told, with a pistol at his head, if he made any noise or exposed them, they would blow his brains out. So effectually was he frightened that he made no complaint, until his friends forced him. Yesterday a warrant was issued, but the culprits had disappeared.

This efficient city officer, Jack Williams, had the common reputation of being a burglar, a highwayman, and a desperado. It was said that he had several times drawn his revolver and levied money contributions on citizens at dead of night in the public streets of Virginia.

Five months after the above item appeared, Williams was assassinated while sitting at a card table one night; a gun was thrust through the crack of the door and Williams dropped from his chair riddled with balls. It was said, at the time, that Williams had been for some time aware that a party

of his own sort (desperadoes) had sworn away his
life; and it was generally believed among the people
that Williams' friends and enemies would make the
assassination memorable — and useful, too — by a
wholesale destruction of each other.*

It did not so happen, but still, times were not dull
during the next twenty-four hours, for within that
time a woman was killed by a pistol shot, a man was
brained with a slung shot, and a man named Reeder
was also disposed of permanently. Some matters in
the *Enterprise* account of the killing of Reeder are
worth noting — especially the accommodating com-

---

* However, one prophecy was verified, at any rate. It was asserted
by the desperadoes that one of their brethren (Joe McGee, *a special
policeman*) was known to be the conspirator chosen by lot to assas-
sinate Williams ; and they also asserted that doom had been pronounced
against McGee, and that he would be assassinated in exactly the same
manner that had been adopted for the destruction of Williams — a
prophecy which came true a year later. After twelve months of distress
(for McGee saw a fancied assassin in every man that approached him),
he made the last of many efforts to get out of the country unwatched.
He went to Carson and sat down in a saloon to wait for the stage — it
would leave at four in the morning. But as the night waned and the
crowd thinned, he grew uneasy, and told the bar-keeper that assassins
were on his track. The bar-keeper told him to stay in the middle of the
room, then, and not go near the door, or the window by the stove. But a
fatal fascination seduced him to the neighborhood of the stove every now
and then, and repeatedly the bar-keeper brought him back to the middle
of the room and warned him to remain there. But he could not. At
three in the morning he again returned to the stove and sat down by a
stranger. Before the bar-keeper could get to him with another warning
whisper, some one outside fired through the window and riddled McGee's
breast with slugs, killing him almost instantly. By the same discharge
the stranger at McGee's side also received attentions which proved fatal
in the course of two or three days.

F **

plaisance of a Virginia justice of the peace. The italics in the following narrative are mine:

MORE CUTTING AND SHOOTING.—The devil seems to have again broken loose in our town. Pistols and guns explode and knives gleam in our streets as in early times. When there has been a long season of quiet, people are slow to wet their hands in blood ; but once blood is spilled, cutting and shooting come easy. Night before last Jack Williams was assassinated, and yesterday forenoon we had more bloody work, growing out of the killing of Williams, and on the same street in which he met his death. It appears that Tom Reeder, a friend of Williams, and George Gumbert were talking, at the meat market of the latter, about the killing of Williams the previous night, when Reeder said it was a most cowardly act to shoot a man in such a way, giving him " no show." Gumbert said that Williams had " as good a show as he gave Billy Brown," meaning the man killed by Williams last March. Reeder said it was a d—d lie, that Williams had no show at all. At this, Gumbert drew a knife and stabbed Reeder, cutting him in two places in the back. One stroke of the knife cut into the sleeve of Reeder's coat and passed downward in a slanting direction through his clothing, and entered his body at the small of the back ; another blow struck more squarely, and made a much more dangerous wound. Gumbert gave himself up to the officers of justice, and was shortly after discharged by Justice Atwill, *on his own recognizance,* to appear for trial at six o'clock in the evening. In the meantime Reeder had been taken into the office of Dr. Owens, where his wounds were properly dressed. *One of his wounds was considered quite dangerous, and it was thought by many that it would prove fatal. But being considerably under the influence of liquor, Reeder did not feel his wounds as he otherwise would, and he got up and went into the street.* He went to the meat market and renewed his quarrel with Gumbert, threatening his life. Friends tried to interfere to put a stop to the quarrel and get the parties away from each other. In the Fashion Saloon Reeder made threats against the life of Gumbert, saying he would kill him, and it is said that *he requested the officers not to arrest Gumbert as he intended to kill him.* After these threats Gumbert went off and procured a double-barreled shot gun, loaded with buck-shot or revolver balls, and went after Reeder. Two or three persons were assisting him along the street, trying to get him home, and had him just in front of the store of Klopstock & Harris, when Gumbert came across toward him from the opposite side of the street with his gun. He came up within about ten or fifteen feet of Reeder, and called out to those with him to "look out! get out of the way!" and they had only time to heed the warning, when he fired. Reeder was at the time attempting to screen himself behind a large cask, which stood against the awning post of Klopstock & Harris's store, but some of the balls took effect in the lower part of his breast, and he reeled around forward and fell in front of the cask. Gumbert then raised his gun and fired the second barrel, which missed Reeder and entered the ground. At the time that this occurred, there were a great many persons on the street in the vicinity, and a number of them called out to Gumbert, when they saw him raise his gun, to "hold on," and "don't shoot!" The cutting took place about ten o'clock and the shooting about twelve. After the shooting the street was instantly crowded with the inhabitants of that part of the town, some appearing much excited and laughing—declaring that it looked like the "good old times of '60." Marshal Perry and officer Birdsall were near when the shooting occurred, and Gumbert was immediately arrested and his gun taken from him, when he was marched off to jail. Many persons who were attracted to the spot where this bloody work had just taken place, looked bewildered and seemed to be asking themselves what was to happen next, appearing in doubt as to whether the killing mania had reached its climax, or whether we were to turn in and have a grand killing spell, shooting whoever might have given us offence. It was whispered around that it was not all over yet—five or six more were to be killed before night. Reeder was taken to the Virginia City Hotel, and doctors called in to examine his wounds. They found that two or three balls had entered his right side; one of them appeared to have passed through the substance of the lungs, while another passed into the liver. Two balls were also found to have struck one of his legs. As some of the balls struck the cask, the wounds in Reeder's leg were probably from these, glancing downwards, though they might have been caused by the second shot fired. After being shot, Reeder said when he got on his feet—smiling as he spoke—"It will take better shooting than that to kill me." The doctors consider it almost impossible for him to recover, but as he has an excellent constitution he may survive, notwithstanding the number and dangerous character of the wounds he has received. The town appears to be perfectly quiet at present, as though the late stormy times had cleared our moral atmosphere; but who can tell in what quarter clouds are lowering or plots ripening?

Reeder — or at least what was left of him — survived his wounds two days! Nothing was ever done
with Gumbert.

Trial by jury is the palladium of our liberties. I
do not know what a palladium is, having never seen
a palladium, but it is a good thing, no doubt, at any
rate. Not less than a hundred men have been
murdered in Nevada — perhaps I would be within
bounds if I said three hundred — and as far as I can
learn, only two persons have suffered the death
penalty there. However, four or five who had no
money and no political influence have been punished
by imprisonment — one languished in prison as
much as eight months, I think. However, I do not
desire to be extravagant — it may have been less.

CHAPTER IX.

THESE murder and jury statistics remind me of
a certain very extraordinary trial and execution
of twenty years ago; it is a scrap of history familiar
to all old Californians, and worthy to be known by
other peoples of the earth that love simple, straight-
forward justice unencumbered with nonsense. I
would apologize for this digression but for the fact
that the information I am about to offer is apology
enough in itself. And since I digress constantly,
anyhow, perhaps it is as well to eschew apologies
altogether and thus prevent their growing irksome.

Capt. Ned Blakely — that name will answer as well
as any other fictitious one (for he was still with the
living at last accounts, and may not desire to be
famous) — sailed ships out of the harbor of San
Francisco for many years. He was a stalwart,
warm-hearted, eagle-eyed veteran, who had been a
sailor nearly fifty years — a sailor from early boyhood. He was a rough, honest creature, full of
pluck, and just as full of hard-headed simplicity, too.
He hated trifling conventionalities — " business "
was the word, with him. He had all a sailor's vin-
dictiveness against the quips and quirks of the law,

and steadfastly believed that the first and last aim
and object of the law and lawyers was to defeat
justice.

He sailed for the Chincha Islands in command of
a guano ship. He had a fine crew, but his negro
mate was his pet — on him he had for years lavished
his admiration and esteem. It was Capt. Ned's first
voyage to the Chinchas, but his fame had gone before him — the fame of being a man who would fight
at the dropping of a handkerchief, when imposed
upon, and would stand no nonsense. It was a fame
well earned. Arrived in the islands, he found that
the staple of conversation was the exploits of one Bill
Noakes, a bully, the mate of a trading ship. This
man had created a small reign of terror there. At
nine o'clock at night, Capt. Ned, all alone, was pac-
ing his deck in the starlight. A form ascended the
side, and approached him. Capt. Ned said:

"Who goes there?"

"I'm Bill Noakes, the best man on the islands."

"What do you want aboard this ship?"

"I've heard of Capt. Ned Blakely, and one of us
is a better man than 'tother — I'll know which,
before I go ashore."

"You have come to the right shop — I'm your
man. I'll learn you to come aboard this ship without an *in*vite."

He seized Noakes, backed him against the main-
mast, pounded his face to a pulp, and then threw
him overboard.

Noakes was not convinced. He returned the next
night, got the pulp renewed, and went overboard head
first, as before. He was satisfied.

A week after this, while Noakes was carousing
with a sailor crowd on shore, at noonday, Capt.
Ned's colored mate came along, and Noakes tried
to pick a quarrel with him. The negro evaded the
trap, and tried to get away. Noakes followed him
up; the negro began to run; Noakes fired on him
with a revolver and killed him. Half a dozen sea-
captains witnessed the whole affair. Noakes retreated to the small after-cabin of his ship, with two
other bullies, and gave out that death would be the
portion of any man that intruded there. There was
no attempt made to follow the villains; there was no
disposition to do it, and indeed very little thought of
such an enterprise. There were no courts and no
officers; there was no government; the islands belonged to Peru, and Peru was far away; she had no
official representative on the ground; and neither
had any other nation.

However, Capt. Ned was not perplexing his head
about such things. They concerned him not. He
was boiling with rage and furious for justice. At
nine o'clock at night he loaded a double-barreled
gun with slugs, fished out a pair of handcuffs, got a
ship's lantern, summoned his quartermaster, and
went ashore. He said:

" Do you see that ship there at the dock?"

" Ay-ay, sir."

" It's the Venus."

" Ay-ay, sir."

" You — you know *me*."

" Ay-ay, sir."

" Very well, then. Take the lantern. Carry it just under your chin. I'll walk behind you and rest this gun-barrel on your shoulder, p'inting forward — so. Keep your lantern well up, so's I can see things ahead of you good. I'm going to march in on Noakes — and take him — and jug the other chaps. If you flinch — well, you know *me*."

" Ay-ay, sir."

In this order they filed aboard softly, arrived at Noakes's den, the quartermaster pushed the door open, and the lantern revealed the three desperadoes sitting on the floor. Capt. Ned said:

" I'm Ned Blakely. I've got you under fire. Don't you move without orders — any of you. You two kneel down in the corner; faces to the wall — now. Bill Noakes, put these handcuffs on; now come up close. Quartermaster, fasten 'em. All right. Don't stir, sir. Quartermaster, put the key in the outside of the door. Now, men, I'm going to lock you two in; and if you try to burst through this door — well, you've heard of *me*. Bill Noakes, fall in ahead, and march. All set. Quartermaster, lock the door."

Noakes spent the night on board Blakely's ship, a prisoner under strict guard. Early in the morning Capt. Ned called in all the sea-captains in the harbor and invited them, with nautical ceremony, to be present on board his ship at nine o'clock to witness the hanging of Noakes at the yard-arm!

"What! The man has not been tried."

"Of course he hasn't. But didn't he kill the nigger?"

"Certainly he did; but you are not thinking of hanging him without a trial?"

"*Trial!* What do I want to try him for, if he killed the nigger?"

"Oh, Capt. Ned, this will *never* do. Think how it will sound."

"Sound be hanged! *Didn't he kill the nigger?*"

"Certainly, certainly, Capt. Ned, — nobody denies that,—but—"

"Then I'm going to *hang* him, that's all. Everybody I've talked to talks just the same way you do. Everybody says he killed the nigger, everybody knows he killed the nigger, and yet every lubber of you wants him *tried* for it. I don't understand such bloody foolishness as that. *Tried!* Mind you, I don't object to trying him if it's got to be done to give satisfaction; and I'll be there, and chip in and help, too; but put it off till afternoon — put it off till afternoon, for I'll have my hands middling full till after the burying—"

"Why, what do you mean? Are you going to hang him *any how*—and try him afterward?"

"Didn't I *say* I was going to hang him? I never saw such people as you. What's the difference?

You ask a favor, and then you ain't satisfied when
you get it. Before or after's all one — *you* know
how the trial will go. He killed the nigger. Say —
I must be going. If your mate would like to come
to the hanging, fetch him along. I like him."

There was a stir in the camp. The captains came
in a body and pleaded with Capt. Ned not to do
this rash thing. They promised that they would
create a court composed of captains of the best
character; they would empanel a jury; they would
conduct everything in a way becoming the serious
nature of the business in hand, and give the case an
impartial hearing and the accused a fair trial. And
they said it would be murder, and punishable by
the American courts if he persisted and hung the
accused on his ship. They pleaded hard. Capt.
Ned said:

"Gentlemen, I'm not stubborn and I'm not un-
reasonable. I'm always willing to do just as near
right as I can. How long will it take?"

"Probably only a little while."

"And can I take him up the shore and hang him
as soon as you are done?"

"If he is proven guilty he shall be hanged without unnecessary delay."

"*If* he's proven guilty. Great Neptune, *ain't*
he guilty? This beats my time. Why you all *know*
he's guilty."

But at last they satisfied him that they were projecting nothing underhanded. Then he said:

"Well, all right. You go on and try him and I'll go down and overhaul his conscience and prepare him to go — like enough he needs it, and I don't want to send him off without a show for hereafter."

This was another obstacle. They finally convinced him that it was necessary to have the accused in court. Then they said they would send a guard to bring him.

"No, sir, I prefer to fetch him myself — he don't get out of *my* hands. Besides, I've got to go to the ship to get a rope, anyway."

The court assembled with due ceremony, empaneled a jury, and presently Capt. Ned entered, leading the prisoner with one hand and carrying a Bible and a rope in the other. He seated himself by the side of his captive and told the court to "up anchor and make sail." Then he turned a searching eye on the jury, and detected Noakes' friends, the two bullies. He strode over and said to them confidentially:

"You're here to interfere, you see. Now you vote right, do you hear?— or else there'll be a double-barreled inquest here when this trial's off, and your remainders will go home in a couple of baskets."

The caution was not without fruit. The jury was a unit — the verdict, "Guilty."

Capt. Ned sprung to his feet and said:

"Come along — you're my meat *now*, my lad,

anyway. Gentlemen you've done yourselves proud.
I invite you all to come and see that I do it all
straight. Follow me to the canyon, a mile above
here."

The court informed him that a sheriff had been
appointed to do the hanging, and—

Capt. Ned's patience was at an end. His wrath
was boundless. The subject of a sheriff was judi-
ciously dropped.

When the crowd arrived at the canyon, Capt. Ned
climbed a tree and arranged the halter, then came
down and noosed his man. He opened his Bible,
and laid aside his hat. Selecting a chapter at ran-
dom, he read it through, in a deep bass voice and
with sincere solemnity. Then he said:

"Lad, you are about to go aloft and give an account of yourself; and the lighter a man's manifest
is, as far as sin's concerned, the better for him.
Make a clean breast, man, and carry a log with
you that'll bear inspection. You killed the nigger?"

No reply. A long pause.

The captain read another chapter, pausing, from
time to time, to impress the effect. Then he talked
an earnest, persuasive sermon to him, and ended
by repeating the question:

"Did you kill the nigger?"

No reply — other than a malignant scowl. The
captain now read the first and second chapters of
Genesis, with deep feeling, paused a moment, closed

the book reverently, and said with a perceptible savor of satisfaction:

"There. Four chapters. There's few that would have took the pains with you that I have."

Then he swung up the condemned, and made the rope fast; stood by and timed him half an hour with his watch, and then delivered the body to the court. A little after, as he stood contemplating the motionless figure, a doubt came into his face; evidently he felt a twinge of conscience — a misgiving — and he said with a sigh:

"Well, p'raps I ought to burnt him, maybe. But I was trying to do for the best."

When the history of this affair reached California (it was in the "early days") it made a deal of talk, but did not diminish the captain's popularity in any degree. It increased it, indeed. California had a population then that "inflicted" justice after a fashion that was simplicity and primitiveness itself, and could therefore admire appreciatively when the same fashion was followed elsewhere.

7••

CHAPTER X.

VICE flourished luxuriantly during the heyday of our "flush times." The saloons were overburdened with custom; so were the police courts, the gambling dens, the brothels, and the jails — unfailing signs of high prosperity in a mining region — in any region, for that matter. Is it not so? A crowded police court docket is the surest of all signs that trade is brisk and money plenty. Still, there is one other sign; it comes last, but when it does come it establishes beyond cavil that the "flush times" are at the flood. This is the birth of the "literary" paper. The *Weekly Occidental*, "devoted to literature," made its appearance in Virginia. All the literary people were engaged to write for it. Mr. F. was to edit it. He was a felicitous skirmisher with a pen, and a man who could say happy things in a crisp, neat way. Once, while editor of the *Union*, he had disposed of a labored, incoherent, two-column attack made upon him by a contemporary, with a single line, which, at first glance, seemed to contain a solemn and tremendous compliment — viz.: "THE LOGIC OF OUR ADVERSARY RESEMBLES THE

PEACE OF GOD,"—and left it to the reader's memory
and afterthought to invest the remark with another
and "more different" meaning by supplying for
himself and at his own leisure the rest of the Scrip-
ture—"*in that it passeth understanding.*" He once
said of a little, half-starved, wayside community that
had no subsistence except what they could get by
preying upon chance passengers who stopped over
with them a day when traveling by the Overland
stage, that in their Church service they had altered
the Lord's Prayer to read: "Give us this day our
daily stranger!"

We expected great things of the *Occidental.* Of
course it could not get along without an original
novel, and so we made arrangements to hurl into the
work the full strength of the company. Mrs. F.
was an able romancist of the ineffable school—I
know no other name to apply to a school whose
heroes are all dainty and all perfect. She wrote the
opening chapter, and introduced a lovely blonde
simpleton who talked nothing but pearls and poetry
and who was virtuous to the verge of eccentricity.
She also introduced a young French Duke of aggravated refinement, in love with the blonde. Mr. F.
followed next week, with a brilliant lawyer who set
about getting the Duke's estates into trouble, and a
sparkling young lady of high society who fell to
fascinating the Duke and impairing the appetite of
the blonde. Mr. D., a dark and bloody editor of
one of the dailies, followed Mr. F., the third week,

7 **

introducing a mysterious Roscicrucian who trans-
muted metals, held consultations with the devil in a
cave at dead of night, and cast the horoscope of the
several heroes and heroines in such a way as to provide plenty of trouble for their future careers and
breed a solemn and awful public interest in the
novel. He also introduced a cloaked and masked
melodramatic miscreant, put him on a salary and set
him on the midnight track of the Duke with a poisoned dagger. He also created an Irish coachman
with a rich brogue and placed him in the service of
the society-young-lady with an ulterior mission to
carry billet-doux to the Duke.

About this time there arrived in Virginia a dissolute stranger with a literary turn of mind — rather
seedy he was, but very quiet and unassuming;
almost diffident, indeed. He was so gentle, and
his manners were so pleasing and kindly, whether he
was sober or intoxicated, that he made friends of all
who came into contact with him. He applied for
literary work, offered conclusive evidence that he
wielded an easy and practiced pen, and so Mr. F.
engaged him at once to help write the novel. His
chapter was to follow Mr. D.'s, and mine was to
come next. Now what does this fellow do but go
off and get drunk and then proceed to his quarters
and set to work with his imagination in a state of
chaos, and that chaos in a condition of extravagant
activity. The result may be guessed. He scanned
the chapters of his predecessors, found plenty of

heroes and heroines already created, and was satisfied with them; he decided to introduce no more;
with all the confidence that whisky inspires and all
the easy complacency it gives to its servant, he then
launched himself lovingly into his work: he married
the coachman to the society-young-lady for the sake
of scandal; married the Duke to the blonde's step-
mother, for the sake of the sensation; stopped the
desperado's salary; created a misunderstanding between the devil and the Roscicrucian; threw the
Duke's property into the wicked lawyer's hands;
made the lawyer's upbraiding conscience drive him
to drink, thence to *delirium tremens*, thence to
suicide; broke the coachman's neck; let his widow
succumb to contumely, neglect, poverty, and consumption; caused the blonde to drown herself, leaving her clothes on the bank with the customary note
pinned to them forgiving the Duke and hoping he
would be happy; revealed to the Duke, by means
of the usual strawberry mark on left arm, that he had
married his own long-lost mother and destroyed his
long-lost sister; instituted the proper and necessary
suicide of the Duke and the Duchess in order to
compass poetical justice; opened the earth and let
the Roscicrucian through, accompanied with the
accustomed smoke and thunder and smell of brim-
stone, and finished with the promise that in the next
chapter, after holding a general inquest, he would
take up the surviving character of the novel and tell
what became of the devil!

G **

It read with singular smoothness, and with a "dead" earnestness that was funny enough to suffocate a body. But there was war when it came in. The other novelists were furious. The mild stranger, not yet more than half sober, stood there, under a scathing fire of vituperation, meek and bewildered, looking from one to another of his assailants, and wondering what he could have done to invoke such a storm. When a lull came at last, he said his say gently and appealingly — said he did not rightly remember what he had written, but was sure he had tried to do the best he could, and knew his object had been to make the novel not only pleasant and plausible but instructive and—

The bombardment began again. The novelists assailed his ill-chosen adjectives and demolished them with a storm of denunciation and ridicule. And so the siege went on. Every time the stranger tried to appease the enemy he only made matters worse. Finally he offered to rewrite the chapter. This arrested hostilities. The indignation gradually quieted down, peace reigned again, and the sufferer retired in safety and got him to his own citadel.

But on the way thither the evil angel tempted him, and he got drunk again. And again his imagination went mad. He led the heroes and heroines a wilder dance than ever; and yet all through it ran that same convincing air of honesty and earnestness that had marked his first work. He got the characters into the most extraordinary situations, put them

through the most surprising performances, and made
them talk the strangest talk! But the chapter cannot be described. It was symmetrically crazy; it
was artistically absurd ; and it had explanatory
footnotes that were fully as curious as the text. I
remember one of the " situations," and will offer it
as an example of the whole. He altered the character of the brilliant lawyer, and made him a great-
hearted, splendid fellow; gave him fame and riches,
and set his age at thirty-three years. Then he made
the blonde discover, through the help of the Roscicrucian and the melodramatic miscreant, that while
the Duke loved her money ardently and wanted it,
he secretly felt a sort of leaning toward the society-
young-lady. Stung to the quick, she tore her affec-
tions from him and bestowed them with tenfold power
upon the lawyer, who responded with consuming
zeal. But the parents would none of it. What they
wanted in the family was a Duke; and a Duke they
were determined to have; though they confessed
that next to the Duke the lawyer had their prefer-
ence. Necessarily, the blonde now went into a decline. The parents were alarmed. They pleaded
with her to marry the Duke, but she steadfastly
refused, and pined on. Then they laid a plan.
They told her to wait a year and a day, and if at
the end of that time she still felt that she could not
marry the Duke, she might marry the lawyer with
their full consent. The result was as they had foreseen: gladness came again, and the flush of return-

ing health.   Then the parents took the next step in
their scheme.   They had the family physician recom-
mend a long sea voyage and much land travel for the
thorough restoration of the blonde's strength; and
they invited the Duke to be of the party.   They
judged that the Duke's constant presence and the
lawyer's protracted absence would do the rest — for
they did not invite the lawyer.

So they set sail in a steamer for America — and
the third day out, when their sea-sickness called
truce and permitted them to take their first meal at
the public table, behold there sat the lawyer!   The
Duke and party made the best of an awkward situation; the voyage progressed, and the vessel neared
America.   But, by and by, two hundred miles off
New Bedford, the ship took fire; she burned to
the water's edge; of all her crew and passengers,
only thirty were saved.   They floated about the sea
half an afternoon and all night long.   Among them
were our friends.   The lawyer, by superhuman ex-
ertions, had saved the blonde and her parents, swimming back and forth two hundred yards and bringing
one each time — (the girl first).   The Duke had
saved himself.   In the morning two whale - ships
arrived on the scene and sent their boats.   The
weather was stormy and the embarkation was attended with much confusion and excitement.   The
lawyer did his duty like a man; helped his exhausted
and insensible blonde, her parents and some others
into a boat (the Duke helped himself in); then a

child fell overboard at the other end of the raft and
the lawyer rushed thither and helped half a dozen
people fish it out, under the stimulus of its mother's
screams. Then he ran back — a few seconds too
late — the blonde's boat was under way. So he had
to take the other boat, and go to the other ship.
The storm increased and drove the vessels out of
sight of each other — drove them whither it would.
When it calmed, at the end of three days, the
blonde's ship was seven hundred miles north of Boston and the other about seven hundred south of that
port. The blonde's captain was bound on a whaling
cruise in the North Atlantic and could not go back
such a distance or make a port without orders; such
being nautical law. The lawyer's captain was to
cruise in the North Pacific, and *he* could not go back
or make a port without orders. All the lawyer's
money and baggage were in the blonde's boat and
went to the blonde's ship — so his captain made him
work his passage as a common sailor. When both
ships had been cruising nearly a year, the one was
off the coast of Greenland and the other in Behring's
Strait. The blonde had long ago been well-nigh
persuaded that her lawyer had been washed over
board and lost just before the whale-ships reached
the raft, and now, under the pleadings of her parents
and the Duke she was at last beginning to nerve herself for the doom of the covenant, and prepare for
the hated marriage. But she would not yield a day
before the date set. The weeks dragged on, the

time narrowed, orders were given to deck the ship for the wedding — a wedding at sea among icebergs and walruses. Five days more, and all would be over. So the blonde reflected, with a sigh and a tear. Oh where was her true love — and why, why did he not come and save her? At that moment he was lifting his harpoon to strike a whale in Behring's Strait, five thousand miles away, by the way of the Arctic Ocean, or twenty thousand by the way of the Horn — that was the reason. He struck, but not with perfect aim — his foot slipped and he fell in the whale's mouth and went down his throat. He was insensible five days. Then he came to himself and heard voices; daylight was streaming through a hole cut in the whale's roof. He climbed out and astonished the sailors who were hoisting blubber up a ship's side. He recognized the vessel, flew aboard, surprised the wedding party at the altar and exclaimed:

"Stop the proceedings — I'm here! Come to my arms, my own!"

There were footnotes to this extravagant piece of literature wherein the author endeavored to show that the whole thing was within the possibilities; he said he got the incident of the whale traveling from Behring's Strait to the coast of Greenland, five thousand miles in five days, through the Arctic Ocean, from Charles Reade's "Love Me Little Love Me Long," and considered that that established the fact that the thing could be done; and he instanced

Jonah's adventure as proof that a man could live in a whale's belly, and added that if a preacher could stand it three days a lawyer could surely stand it five!

There was a fiercer storm than ever in the editorial sanctum now, and the stranger was peremptorily discharged, and his manuscript flung at his head. But he had already delayed things so much that there was not time for some one else to rewrite the chapter, and so the paper came out without any novel in it. It was but a feeble, struggling, stupid journal, and the absence of the novel probably shook public confidence; at any rate, before the first side of the next issue went to press, the *Weekly Occidental* died as peacefully as an infant.

An effort was made to resurrect it, with the proposed advantage of a telling new title, and Mr. F. said that *The Phenix* would be just the name for it, because it would give the idea of a resurrection from its dead ashes in a new and undreamed of condition of splendor; but some low-priced smarty on one of the dailies suggested that we call it the *Lazarus;* and inasmuch as the people were not profound in Scriptural matters, but thought the resurrected Lazarus and the dilapidated mendicant that begged in the rich man's gateway were one and the same person, the name became the laughing-stock of the town, and killed the paper for good and all.

I was sorry enough, for I was very proud of being connected with a literary paper — prouder than I

have ever been of anything since, perhaps. I had
written some rhymes for it — poetry I considered it
— and it was a great grief to me that the production
was on the "first side" of the issue that was not
completed, and hence did not see the light. But
time brings its revenges — I can put it in here; it
will answer in place of a tear dropped to the memory
of the lost *Occidental*. The idea (not the chief idea,
but the vehicle that bears it) was probably suggested by the old song called "The Raging Canal,"
but I cannot remember now. I do remember,
though, that at that time I thought my doggerel
was one of the ablest poems of the age:

THE AGED PILOT MAN

On the Erie Canal, it was,
　　All on a summer's day,
I sailed forth with my parents
　　Far away to Albany.

From out the clouds at noon that day
　　There came a dreadful storm,
That piled the billows high about,
　　And filled us with alarm.

A man came rushing from a house,
　　Saying, " Snub up * your boat, I pray,
Snub up your boat, snub up, alas,
　　Snub up while yet you may."

Our captain cast one glance astern,
　　Then forward glancèd he,
And said, " My wife and little ones
　　I never more shall see."

---

* The customary canal technicality for " tie up."

Said Dollinger the pilot man,
   In noble words, but few,—
"Fear not, but lean on Dollinger,
   And he will fetch you through."

The boat drove on, the frightened mules
   Tore through the rain and wind,
And bravely still, in danger's post,
   The whip-boy strode behind.

"Come 'board, come 'board," the captain cried,
   "Nor tempt so wild a storm;"
But still the raging mules advanced,
   And still the boy strode on.

Then said the captain to us all,
   "Alas, 'tis plain to me,
The greater danger is not there,
   But here upon the sea.

"So let us strive, while life remains,
   To save all souls on board,
And then if die at last we must,
   Let . . . . I *cannot* speak the word!"

Said Dollinger the pilot man,
   Tow'ring above the crew,
"Fear not, but trust in Dollinger,
   And he will fetch you through."

"Low bridge! low bridge!" all heads went down,
   The laboring bark sped on;
A mill we passed, we passed a church,
   Hamlets, and fields of corn;
And all the world came out to see,
   And chased along the shore

Crying, "Alas, alas, the sheeted rain,
   The wind, the tempest's roar!
Alas, the gallant ship and crew,
   Can *nothing* help them more?"

And from our deck sad eyes looked out
   Across the stormy scene;
The tossing wake of billows aft,
   The bending forests green,

The chickens sheltered under carts,
   In lee of barn the cows,
The skurrying swine with straw in mouth,
   The wild spray from our bows!

     " She balances!
      She wavers!
Now let her go about!
If she misses stays and broaches to,
We're all " — [then with a shout]
    " Huray! huray!
      Avast! belay!
      Take in more sail!
      Lord, what a gale!
Ho, boy, haul taut on the hind mule's tail!

" Ho! lighten ship! ho! man the pump!
   Ho, hostler, heave the lead!
And count ye all, both great and small,
   As numbered with the dead!
For mariner for forty year
   On Erie, boy and man,
I never yet saw such a storm,
   Or one 't with it began! "

So overboard a keg of nails
   And anvils three we threw,
Likewise four bales of gunny-sacks,
   Two hundred pounds of glue,
Two sacks of corn, four ditto wheat,
   A box of books, a cow,
A violin, Lord Byron's works,
   A rip-saw and a sow.

A curve! a curve! the dangers grow!
" Labbord ! — stabbord ! — s-t-e-a-d-y ! — so ! —
Hard-a-port, Dol! — hellum-a-lee !
Haw the head mule ! — the aft one gee !
Luff ! — bring her to the wind ! "

" A quarter-three ! — 'tis shoaling fast !
   Three feet large ! — t-h-r-e-e feet ! —
Three feet scant ! " I cried in fright
   " Oh, is there *no* retreat?"

Said Dollinger the pilot man,
   As on the vessel flew,
" Fear not, but trust in Dollinger,
   And he will fetch you through."

A panic struck the bravest hearts,
   The boldest cheek turned pale;
For plain to all, this shoaling said
A leak had burst the ditch's bed !
And, straight as bolt from crossbow sped,
Our ship swept on with shoaling lead,
   Before the fearful gale !

" Sever the tow line ! Cripple the mules ! "
   Too late ! . . . . . There comes a shock !
   .   .   .   .   .
Another length, and the fated craft
   Would have swum in the saving lock !

Then gathered together the shipwrecked crew
   And took one last embrace,
While sorrowful tears from despairing eyes
   Ran down each hopeless face;
And some did think of their little ones
   Whom they never more might see,
And others of waiting wives at home,
   And mothers that grieved would be.

But of all the children of misery there
  On that poor sinking frame,
But one spake words of hope and faith,
  And I worshipped as they came:
Said Dollinger the pilot man,—
  (O brave heart, strong and true!)—
" Fear not, but trust in Dollinger,
  For he will fetch you through."

Lo! scarce the words have passed his lips
  The dauntless prophet say'th,
When every soul about him seeth
  A wonder crown his faith!

For straight a farmer brought a plank,—
  (Mysteriously inspired)—
And laying it unto the ship,
  In silent awe retired.
Then every sufferer stood amazed
  That pilot man before;
A moment stood. Then wondering turned,
  And speechless walked ashore.

CHAPTER XI.

SINCE I desire, in this chapter, to say an instructive word or two about the silver mines, the
reader may take this fair warning and skip, if he
chooses. The year 1863 was perhaps the very top
blossom and culmination of the "flush times."
Virginia swarmed with men and vehicles to that
degree that the place looked like a very hive — that
is when one's vision could pierce through the thick
fog of alkali dust that was generally blowing in
summer. I will say, concerning this dust, that if
you drove ten miles through it, you and your horses
would be coated with it a sixteenth of an inch thick
and present an outside appearance that was a uniform pale yellow color, and your buggy would have
three inches of dust in it. thrown there by the
wheels. The delicate scales used by the assayers
were inclosed in glass cases intended to be air-tight,
and yet some of this dust was so impalpable and so
invisibly fine that it would get in, somehow, and
impair the accuracy of those scales.

Speculation ran riot, and yet there was a world of
substantial business going on, too. All freights
were brought over the mountains from California

(150 miles) by pack-train partly, and partly in huge
wagons drawn by such long mule teams that each
team amounted to a procession, and it did seem,
sometimes, that the grand combined procession of
animals stretched unbroken from Virginia to California. Its long route was traceable clear across
the deserts of the territory by the writhing serpent
of dust it lifted up. By these wagons, freights
over that hundred and fifty miles were $200 a ton
for small lots (same price for all express matter
brought by stage), and $100 a ton for full loads.
One Virginia firm received one hundred tons of
freight a month, and paid $10,000 a month freight-
age. In the winter the freights were much higher.
All the bullion was shipped in bars by stage to San
Francisco (a bar was usually about twice the size of
a pig of lead and contained from $1,500 to $3,000,
according to the amount of gold mixed with the
silver), and the freight on it (when the shipment
was large) was one and a quarter per cent. of its
intrinsic value. So, the freight on these bars probably averaged something more than $25 each.
Small shippers paid two per cent. There were
three stages a day, each way, and I have seen the
outgoing stages carry away a third of a ton of
bullion each, and more than once I saw them divide
a two-ton lot and take it off. However, these were
extraordinary events.* Two tons of silver bullion

---

* Mr. Valentine, Wells-Fargo's agent, has handled all the bullion
shipped through the Virginia office for many a month. To his mem-

would be in the neighborhood of forty bars, and the freight on it over $1,000. Each coach always carried a deal of ordinary express matter beside, and also from fifteen to twenty passengers at from $25 to $30 a head. With six stages going all the time, Wells, Fargo & Co.'s Virginia City business was important and lucrative.

All along under the center of Virginia and Gold

---

ory — which is excellent — we are indebted for the following exhibit of the company's business in the Virginia office since the first of January, 1862: From January 1st to April 1st, about $270,000 worth of bullion passed through that office; during the next quarter, $570,000; next quarter, $800,000; next quarter, $956,000; next quarter, $1,275,000; and for the quarter ending on the 30th of last June, about $1,600,000. Thus in a year and a half, the Virginia office only shipped $5,330,000 in bullion. During the year 1862 they shipped $2,615,000, so we perceive the average shipments have more than doubled in the last six months. This gives us room to promise for the Virginia office $500,000 a month for the year 1863 (though perhaps, judging by the steady increase in the business, we are underestimating, somewhat). This gives us $6,000,000 for the year. Gold Hill and Silver City together can beat us — we will give them $10,000,000. To Dayton, Empire City, Ophir, and Carson City, we will allow an aggregate of $8,000,000, which is not over the mark, perhaps, and may possibly be a little under it. To Esmeralda we give $4,000,000. To Reese River and Humboldt $2,000,000, which is liberal now, but may not be before the year is out. So we prognosticate that the yield of bullion this year will be about $30,000,000. Placing the number of mills in the Territory at one hundred, this gives to each the labor of producing $300,000 in bullion during the twelve months. Allowing them to run three hundred days in the year (which none of them more than do), this makes their work average $1,000 a day. Say the mills average twenty tons of rock a day, and this rock worth $50 as a general thing, and you have the actual work of our one hundred mills figured down "to a spot" $1,000 a day each, and $30,000,000 a year in the aggregate.— *Enterprise*.

[A considerable overestimate.— M. T.]

8**

Hill, for a couple of miles, ran the great Comstock
silver lode — a vein of ore from fifty to eighty feet
thick between its solid walls of rock — a vein as
wide as some of New York's streets. I will remind
the reader that in Pennsylvania a coal vein only
eight feet wide is considered ample.

Virginia was a busy city of streets and houses
above ground. Under it was another busy city,
down in the bowels of the earth, where a great
population of men thronged in and out among an
intricate maze of tunnels and drifts, flitting hither
and thither under a winking sparkle of lights, and
over their heads towered a vast web of interlocking
timbers that held the walls of the gutted Comstock
apart. These timbers were as large as a man's
body, and the framework stretched upward so far
that no eye could pierce to its top through the
closing gloom. It was like peering up through the
clean-picked ribs and bones of some colossal skeleton. Imagine such a framework two miles long,
sixty feet wide, and higher than any church spire in
America. Imagine this stately lattice-work stretching down Broadway, from the St. Nicholas to Wall
street, and a Fourth of July procession, reduced to
pigmies, parading on top of it and flaunting their
flags, high above the pinnacle of Trinity steeple.
One can imagine that, but he cannot well imagine
what that forest of timbers cost, from the time they
were felled in the pineries beyond Washoe Lake,
hauled up and around Mount Davidson at atrocious

rates of freightage, then squared, let down into the
deep maw of the mine and built up there. Twenty
ample fortunes would not timber one of the greatest
of those silver mines. The Spanish proverb says it
requires a gold mine to "run" a silver one, and it
is true. A beggar with a silver mine is a pitiable
pauper indeed if he cannot sell.

I spoke of the underground Virginia as a city.
The **Gould & Curry** is only one single mine under
there, among a great many others; yet the Gould
& Curry's streets of dismal drifts and tunnels were
five miles in extent, altogether, and its population
five hundred miners. Taken as a whole, the underground city had some thirty miles of streets and a
population of five or six thousand. In this present
day some of those populations are at work from
twelve to sixteen hundred feet under Virginia and
Gold Hill, and the signal-bells that tell them what
the superintendent above ground desires them to do
are struck by telegraph as we strike a fire alarm.
Sometimes men fall down a shaft, there, a thousand
feet deep. In such cases, the usual plan is to hold
an inquest.

If you wish to visit one of those mines, you may
walk through a tunnel about half a mile long if you
prefer it, or you may take the quicker plan of
shooting like a dart down a shaft, on a small plat-
form. It is like tumbling down through an empty
steeple, feet first. When you reach the bottom,
you take a candle and tramp through drifts and

H **

tunnels where throngs of men are digging and blasting; you watch them send up tubs full of great
lumps of stone — silver ore; you select choice
specimens from the mass, as souvenirs; you admire
the world of skeleton timbering; you reflect frequently that you are buried under a mountain, a
thousand feet below daylight; being in the bottom
of the mine you climb from "gallery" to "gallery," up endless ladders that stand straight up and
down; when your legs fail you at last, you lie down
in a small box-car in a cramped "incline" like a
half up-ended sewer and are dragged up to daylight
feeling as if you are crawling through a coffin that
has no end to it. Arrived at the top, you find a
busy crowd of men receiving the ascending cars and
tubs and dumping the ore from an elevation into
long rows of bins capable of holding half a dozen
tons each; under the bins are rows of wagons load-
ing from chutes and trap-doors in the bins, and down
the long street is a procession of these wagons
wending toward the silver mills with their rich
freight. It is all "done," now, and there you are.
You need never go down again, for you have seen
it all. If you have forgotten the process of re-
ducing the ore in the mill and making the silver
bars, you can go back and find it again in my
Esmeralda chapters, if so disposed.

Of course these mines cave in, in places, occasionally, and then it is worth one's while to take the
risk of descending into them and observing the

crushing power exerted by the pressing weight of a
settling mountain. I published such an experience
in the *Enterprise*, once, and from it I will take an
extract:

AN HOUR IN THE CAVED MINES.— We journeyed down into the
Ophir mine, yesterday, to see the earthquake. We could not go down
the deep incline, because it still has a propensity to cave in places.
Therefore we traveled through the long tunnel which enters the hill
above the Ophir office, and then by means of a series of long ladders,
climbed away down from the first to the fourth gallery. Traversing a
drift, we came to the Spanish line, passed five sets of timbers still un-
injured, and found the earthquake. Here was as complete a chaos as
ever was seen — vast masses of earth and splintered and broken timbers
piled confusedly together, with scarcely an aperture left large enough
for a cat to creep through. Rubbish was still falling at intervals from
above, and one timber which had braced others earlier in the day, was
now crushed down out of its former position, showing that the caving
and settling of the tremendous mass was still going on. We were in
that portion of the Ophir known as the "north mines." Returning to
the surface, we entered a tunnel leading into the Central, for the purpose of getting into the main Ophir. Descending a long incline in this
tunnel, we traversed a drift or so, and then went down a deep shaft
from whence we proceeded into the fifth gallery of the Ophir. From a
side-drift we crawled through a small hole and got into the midst of the
earthquake again — earth and broken timbers mingled together without
regard to grace or symmetry. A large portion of the second, third, and
fourth galleries had caved in and gone to destruction — the two latter at
seven o'clock on the previous evening.

At the turn-table, near the northern extremity of the fifth gallery,
two big piles of rubbish had forced their way through from the fifth
gallery, and from the looks of the timbers, more was about to come.
These beams are solid — eighteen inches square; first, a great beam is
laid on the floor, then upright ones, five feet high, stand on it, support-
ing another horizontal beam, and so on, square above square, like the
framework of a window. The superincumbent weight was sufficient to
mash the ends of those great upright beams fairly into the solid wood of
the horizontal ones three inches, compressing and bending the upright

beam till it curved like a bow. Before the Spanish caved in, some of their twelve-inch horizontal timbers were compressed in this way until they were only five inches thick! Imagine the power it must take to squeeze a solid log together in that way. Here, also, was a range of timbers, for a distance of twenty feet, tilted six inches out of the perpendicular by the weight resting upon them from the caved galleries above. You could hear things cracking and giving way, and it was not pleasant to know that the world overhead was slowly and silently sinking down upon you. The men down in the mine do not mind it, however.

Returning along the fifth gallery, we struck the safe part of the Ophir incline, and went down it to the sixth; but we found ten inches of water there, and had to come back. In repairing the damage done to the incline, the pump had to be stopped for two hours, and in the meantime the water gained about a foot. However, the pump was at work again, and the flood-water was decreasing. We climbed up to the fifth gallery again and sought a deep shaft whereby we might descend to another part of the sixth, out of reach of the water, but suffered disappointment, as the men had gone to dinner, and there was no one to man the windlass. So, having seen the earthquake, we climbed out at the Union incline and tunnel, and adjourned, all dripping with candle grease and perspiration, to lunch at the Ophir office.

During the great flush year of 1863, Nevada [claims to have] produced $25,000,000 in bullion — almost, if not quite, a round million to each thousand inhabitants, which is very well, considering that she was without agriculture and manufactures.* Silver mining was her sole productive industry.

---

* Since the above was in type, I learn from an official source that the above figure is too high, and that the yield for 1863 did not exceed $20,000,000. However, the day for large figures is approaching; the Sutro Tunnel is to plow through the Comstock lode from end to end, at a depth of two thousand feet, and then mining will be easy and comparatively inexpensive; and the momentous matters of drainage and hoisting and hauling of ore will cease to be burdensome. This vast

work will absorb many years, and millions of dollars, in its completion; but it will early yield money, for that desirable epoch will begin as soon as it strikes the first end of the vein. The tunnel will be some eight miles long, and will develop astonishing riches. Cars will carry the ore through the tunnel and dump it in the mills, and thus do away with the present costly system of double handling and transportation by mule teams. The water from the tunnel will furnish the motive power for the mills. Mr. Sutro, the originator of this prodigious enterprise, is one of the few men in the world who is gifted with the pluck and perseverance necessary to follow up and hound such an undertaking to its completion. He has converted several obstinate Congresses to a deserved friendliness toward his important work, and has gone up and down and to and fro in Europe until he has enlisted a great moneyed interest in it there.

CHAPTER XII.

EVERY now and then, in these days, the boys used to tell me I ought to get one Jim Blaine to tell me the stirring story of his grandfather's old ram — but they always added that I must not mention the matter unless Jim was drunk at the time — just comfortably and sociably drunk. They kept this up until my curiosity was on the rack to hear the story. I got to haunting Blaine; but it was of no use, the boys always found fault with his condition; he was often moderately but never satisfactorily drunk. I never watched a man's condition with such absorbing interest, such anxious solicitude; I never so pined to see a man uncompromisingly drunk before. At last, one evening I hurried to his cabin, for I learned that this time his situation was such that even the most fastidious could find no fault with it — he was tranquilly, serenely, symmetrically drunk — not a hiccup to mar his voice, not a cloud upon his brain thick enough to obscure his memory. As I entered, he was sitting upon an empty powder-keg, with a clay pipe in one hand and the other raised to command silence. His face

was round, red, and very serious; his throat was
bare and his hair tumbled; in general appearance
and costume he was a stalwart miner of the period.
On the pine table stood a candle, and its dim light
revealed "the boys" sitting here and there on
bunks, candle-boxes, powder-kegs, etc. They said:

"Sh—! Don't speak—he's going to com-
mence."

THE STORY OF THE OLD RAM

I found a seat at once, and Blaine said:

"I don't reckon them times will ever come again.
There never was a more bullier old ram than what
he was. Grandfather fetched him from Illinois—
got him of a man by the name of Yates—Bill
Yates—maybe you might have heard of him; his
father was a deacon—Baptist—and he was a
rustler, too; a man had to get up ruther early to
get the start of old Thankful Yates; it was him that
put the Greens up to jining teams with my grandfather when he moved west. Seth Green was
prob'ly the pick of the flock; he married a Wilkerson—Sarah Wilkerson—good cretur, she was—
one of the likeliest heifers that was ever raised in
old Stoddard, everybody said that knowed her.
She could heft a bar'l of flour as easy as I can flirt
a flapjack. And spin? Don't mention it! Independent? Humph! When Sile Hawkins come a
browsing around her, she let him know that for all
his tin he couldn't trot in harness alongside of *her*.
You see, Sile Hawkins was—no, it warn't Sile

Hawkins, after all — it was a galoot by the name ol
Filkins — I disremember his first name; but he *was*
a stump — come into pra'r meeting drunk, one
night, hooraying for Nixon, becuz he thought it was
a primary; and old Deacon Ferguson up and
scooted him through the window and he lit on old
Miss Jefferson's head, poor old filly. She was a
good soul — had a glass eye and used to lend it to
old Miss Wagner, that hadn't any, to receive company in; it warn't big enough, and when Miss
Wagner warn't noticing, it would get twisted
around in the socket, and look up, maybe, or out
to one side, and every which way, while t' other
one was looking as straight ahead as a spyglass.
Grown people didn't mind it, but it most always
made the children cry, it was so sort of scary. She
tried packing it in raw cotton, but it wouldn't work,
somehow — the cotton would get loose and stick
out and look so kind of awful that the children
couldn't stand it no way. She was always dropping
it out, and turning up her old deadlight on the
company empty, and making them oncomfortable,
becuz *she* never could tell when it hopped out,
being blind on that side, you see. So somebody
would have to hunch her and say, ' Your game eye
has fetched loose, Miss Wagner, dear' — and then all
of them would have to sit and wait till she jammed
it in again — wrong side before, as a general thing,
and green as a bird's egg, being a bashful cretur
and easy sot back before company. But being

wrong side before warn't much difference, anyway, becuz her own eye was sky-blue and the glass one was yaller on the front side, so whichever way she turned it it didn't match nohow. Old Miss Wagner was considerable on the borrow, she was. When she had a quilting, or Dorcas S'iety at her house she gen'ally borrowed Miss Higgins's wooden leg to stump around on; it was considerable shorter than her other pin, but much *she* minded that. She said she couldn't abide crutches when she had company, becuz they were so slow; said when she had company and things had to be done, she wanted to get up and hump herself. She was as bald as a jug, and so she used to borrow Miss Jacops's wig — Miss Jacops was the coffin-peddler's wife — a ratty old buzzard, he was, that used to go roosting around where people was sick, waiting for 'em; and there that old rip would sit all day, in the shade, on a coffin that he judged would fit the can'idate; and if it was a slow customer and kind of uncertain, he'd fetch his rations and a blanket along and sleep in the coffin nights. He was anchored out that way, in frosty weather, for about three weeks, once, before old Robbins's place, waiting for him; and after that, for as much as two years, Jacops was not on speaking terms with the old man, on account of his disapp'inting him. He got one of his feet froze, and lost money, too, becuz old Robbins took a favorable turn and got well. The next time Robbins got sick, Jacops tried to make up with him,

and varnished up the same old coffin and fetched it
along; but old Robbins was too many for him; he
had him in, and 'peared to be powerful weak; he
bought the coffin for ten dollars and Jacops was to
pay it back and twenty-five more besides if Robbins
didn't like the coffin after he'd tried it. And then
Robbins died, and at the funeral he bursted off the
lid and riz up in his shroud and told the parson to
let up on the performances, becuz he could *not*
stand such a coffin as that. You see he had been
in a trance once before, when he was young, and
he took the chances on another, cal'lating that if he
made the trip it was money in his pocket, and if he
missed fire he couldn't lose a cent. And, by George,
he sued Jacops for the rhino and got judgment; and
he set up the coffin in his back parlor and said he
'lowed to take his time, now. It was always an
aggravation to Jacops, the way that miserable old
thing acted. He moved back to Indiany pretty
soon — went to Wellsville — Wellsville was the
place the Hogadorns was from. Mighty fine family.
Old Maryland stock. Old Squire Hogadorn could
carry around more mixed licker, and cuss better
than most any man I ever see. His second wife
was the Widder Billings — she that was Becky
Martin; her dam was Deacon Dunlap's first wife.
Her oldest child, Maria, married a missionary and
died in grace——et up by the savages. They et
him, too, poor feller — biled him. It warn't the
custom, so they say, but they explained to friends

of his'n that went down there to bring away his
things, that they'd tried missionaries every other
way and never could get any good out of 'em —
and so it annoyed all his relations to find out that
that man's life was fooled away just out of a dern'd
experiment, so to speak. But mind you, there ain't
anything ever reely lost; everything that people
can't understand and don't see the reason of does
good if you only hold on and give it a fair shake;
Prov'dence don't fire no blank ca'tridges, boys.
That there missionary's substance, unbeknowns to
himself, actu'ly converted every last one of them
heathens that took a chance at the barbecue.
Nothing ever fetched them but that. Don't tell *me*
it was an accident that he was biled. There ain't
no such a thing as an accident. When my Uncle
Lem was leaning up agin a scaffolding once, sick,
or drunk, or suthin, an Irishman with a hod full of
bricks fell on him out of the third story and broke
the old man's back in two places. People said it
was an accident. Much accident there was about
that. He didn't know what he was there for, but
he was there for a good object. If he hadn't been
there the Irishman would have been killed. Nobody
can ever make me believe anything different from
that. Uncle Lem's dog was there. Why didn't
the Irishman fall on the dog? Becuz the dog would
a seen him a coming and stood from under. That's
the reason the dog warn't appinted. A dog can't
be depended on to carry out a special providence.

Mark my words, it was a put-up thing. Accidents
don't happen, boys. Uncle Lem's dog — I wish
you could a seen that dog. He was a reglar shepherd — or ruther he was part bull and part shepherd
— splendid animal; belonged to Parson Hagar before Uncle Lem got him. Parson Hagar belonged
to the Western Reserve Hagars; prime family; his
mother was a Watson; one of his sisters married a
Wheeler; they settled in Morgan County, and he
got nipped by the machinery in a carpet factory
and went through in less than a quarter of a minute;
his widder bought the piece of carpet that had his
remains wove in, and people come a hundred mile
to 'tend the funeral. There was fourteen yards in
the piece. She wouldn't let them roll him up, but
planted him just so — full length. The church was
middling small where they preached the funeral,
and they had to let one end of the coffin stick out
of the window. They didn't bury him — they
planted one end, and let him stand up, same as a
monument. And they nailed a sign on it and put —
put on — put on it — sacred to — the m-e-m-o-r-y
— of fourteen y-a-r-d-s — of three-ply — car - - - pet
— containing all that was — m-o-r-t-a-l — of — of —
W-i-l-l-i-a-m — W-h-e —''

Jim Blaine had been growing gradually drowsy
and drowsier — his head nodded, once, twice, three
times — dropped peacefully upon his breast, and he
fell tranquilly asleep. The tears were running down
the boys' cheeks — they were suffocating with sup-

pressed laughter — and had been from the start, though I had never noticed it. I perceived that I was "sold." I learned then that Jim Blaine's peculiarity was that whenever he reached a certain stage of intoxication, no human power could keep him from setting out, with impressive unction, to tell about a wonderful adventure which he had once had with his grandfather's old ram — and the mention of the ram in the first sentence was as far as any man had ever heard him get, concerning it. He always maundered off, interminably, from one thing to another, till his whisky got the best of him, and he fell asleep. What the thing was that happened to him and his grandfather's old ram is a dark mystery to this day, for nobody has ever yet found out.

9••

CHAPTER XIII.

OF course there was a large Chinese population in Virginia — it is the case with every town and city on the Pacific coast. They are a harmless race when white men either let them alone or treat them no worse than dogs; in fact, they are almost entirely harmless anyhow, for they seldom think of resenting the vilest insults or the cruelest injuries. They are quiet, peaceable, tractable, free from drunkenness, and they are as industrious as the day is long. A disorderly Chinaman is rare, and a lazy one does not exist. So long as a Chinaman has strength to use his hands he needs no support from anybody; white men often complain of want of work, but a Chinaman offers no such complaint; he always manages to find something to do. He is a great convenience to everybody — even to the worst class of white men, for he bears the most of their sins, suffering fines for their petty thefts, imprisonment for their robberies, and death for their murders. Any white man can swear a Chinaman's life away in the courts, but no Chinaman can testify against a white man. Ours is the " land of the free "— nobody denies that — nobody challenges it. [Maybe it is because we won't let other people testify.] As I write, news comes that in broad daylight in San Francisco, some boys have stoned an inoffensive Chinaman to death, and that although a large crowd witnessed the shameful deed, no one interfered.

There are seventy thousand (and possibly one hundred thousand) Chinamen on the Pacific coast. There were about a thousand in Virginia. They were penned into a "Chinese quarter"—a thing which they do not particularly object to, as they are fond of herding together. Their buildings were of wood; usually only one story high, and set thickly together along streets scarcely wide enough for a wagon to pass through. Their quarter was a little removed from the rest of the town. The chief employment of Chinamen in towns is to wash clothing. They always send a bill pinned to the clothes. It is mere ceremony, for it does not enlighten the customer much. Their price for washing was $2.50 per dozen — rather cheaper than white people could afford to wash for at that time. A very common sign on the Chinese houses was: "See Yup, Washer and Ironer;" "Hong Wo, Washer;" "Sam Sing & Ah Hop, Washing." The house servants, cooks, etc., in California and Nevada, were chiefly Chinamen. There were few white servants and no Chinawomen so employed. Chinamen make good house servants, being quick, obedient, patient, quick to learn, and tirelessly in-

9**

dustrious. They do not need to be taught a thing twice, as a general thing. They are imitative. If a Chinaman were to see his master break up a center table, in a passion, and kindle a fire with it, that Chinaman would be likely to resort to the furniture for fuel forever afterward.

All Chinamen can read, write, and cipher with easy facility — pity but all our petted *voters* could. In California they rent little patches of ground and do a deal of gardening. They will raise surprising crops of vegetables on a sand pile. They waste nothing. What is rubbish to a Christian, a Chinaman carefully preserves and makes useful in one way or another. He gathers up all the old oyster and sardine cans that white people throw away, and procures marketable tin and solder from them by melting. He gathers up old bones and turns them into manure. In California he gets a living out of old mining claims that white men have abandoned as exhausted and worthless — and then the officers come down on him once a month with an exorbitant swindle to which the legislature has given the broad, general name of " foreign " mining tax, but it is usually inflicted on no foreigners but Chinamen. This swindle has in some cases been repeated once or twice on the same victim in the course of the same month — but the public treasury was not additionally enriched by it, probably.

Chinamen hold their dead in great reverence — they worship their departed ancestors, in fact.

Hence, in China, a man's front yard, back yard, or any other part of his premises, is made his family burying-ground, in order that he may visit the graves at any and all times. Therefore that huge empire is one mighty cemetery; it is ridged and wrinkled from its center to its circumference with graves — and inasmuch as every foot of ground must be made to do its utmost, in China, lest the swarming population suffer for food, the very graves are cultivated and yield a harvest, custom holding this to be no dishonor to the dead. Since the departed are held in such worshipful reverence, a Chinaman cannot bear that any indignity be offered the places where they sleep. Mr. Burlingame said that herein lay China's bitter opposition to railroads; a road could not be built anywhere in the empire without disturbing the graves of their ancestors or friends.

A Chinaman hardly believes he could enjoy the hereafter except his body lay in his beloved China; also, he desires to receive, himself, after death, that worship with which he has honored his dead that preceded him. Therefore, if he visits a foreign country, he makes arrangements to have his bones returned to China in case he dies; if he hires to go to a foreign country on a labor contract, there is always a stipulation that his body shall be taken back to China if he dies; if the government sells a gang of coolies to a foreigner for the usual five-year term, it is specified in the contract that their bodies shall be restored to China in case of death. On the

I**

Pacific coast the Chinamen all belong to one or another of several great companies or organizations, and these companies keep track of their members, register their names, and ship their bodies home when they die. The See Yup Company is held to be the largest of these. The Ning Yeong Company is next, and numbers eighteen thousand members on the coast. Its headquarters are at San Francisco, where it has a costly temple, several great officers (one of whom keeps regal state in seclusion and cannot be approached by common humanity), and a numerous priesthood. In it I was shown a register of its members, with the dead and the date of their shipment to China duly marked. Every ship that sails from San Francisco carries away a heavy freight of Chinese corpses — or did, at least, until the legislature, with an ingenious refinement of Christian cruelty, forbade the shipments, as a neat underhanded way of deterring Chinese immigration. The bill was offered, whether it passed or not. It is my impression that it passed. There was another bill — it became a law — compelling every incoming Chinaman to be vaccinated on the wharf and pay a duly-appointed quack (no decent doctor would defile himself with such legalized robbery) ten dollars for it. As few importers of Chinese would want to go to an expense like that, the lawmakers thought this would be another heavy blow to Chinese immigration.

What the Chinese quarter of Virginia was like —

or, indeed, what the Chinese quarter of any Pacific coast town was and is like — may be gathered from this item which I printed in the *Enterprise* while reporting for that paper:

CHINATOWN.— Accompanied by a fellow reporter, we made a trip through our Chinese quarter the other night. The Chinese have built their portion of the city to suit themselves; and as they keep neither carriages nor wagons, their streets are not wide enough, as a general thing, to admit of the passage of vehicles. At ten o'clock at night the Chinaman may be seen in all his glory. In every little cooped-up, dingy cavern of a hut, faint with the odor of burning Josh-lights and with nothing to see the gloom by save the sickly, guttering tallow candle, were two or three yellow, long-tailed vagabonds, coiled up on a sort of short truckle-bed, smoking opium, motionless and with their lustreless eyes turned inward from excess of satisfaction — or rather the recent smoker looks thus, immediately after having passed the pipe to his neighbor — for opium-smoking is a comfortless operation, and requires constant attention. A lamp sits on the bed, the length of the long pipe-stem from the smoker's mouth; he puts a pellet of opium on the end of a wire, sets it on fire, and plasters it into the pipe much as a Christian would fill a hole with putty; then he applies the bowl to the lamp and proceeds to smoke — and the stewing and frying of the drug and the gurgling of the juices in the stem would wellnigh turn the stomach of a statue. John likes it, though; it soothes him; he takes about two dozen whiffs, and then rolls over to dream, Heaven only knows what, for we could not imagine by looking at the soggy creature. Possibly in his visions he travels far away from the gross world and his regular washing, and feasts on succulent rats and birds'-nests in Paradise.

Mr. Ah Sing keeps a general grocery and provision store at No. 13 Wang street. He lavished his hospitality upon our party in the friendliest way. He had various kinds of colored and colorless wines and brandies, with unpronounceable names, imported from China in little crockery jugs, and which he offered to us in dainty little miniature wash-basins of porcelain. He offered us a mess of birds'-nests; also, small, neat sausages, of which we could have swallowed several yards if we had chosen to try, but we suspected that each link contained the corpse of a mouse, and therefore refrained. Mr. Sing had in his store a thousand articles of merchandise, curious to behold, impossible to imagine the uses of, and beyond our ability to describe.

His ducks, however, and his eggs, we could understand; the former were split open and flattened out like codfish, and came from China in that shape, and the latter were plastered over with some kind of paste which kept them fresh and palatable through the long voyage.

We found Mr. Hong Wo, No. 37 Chow-chow street, making up a lottery scheme — in fact, we found a dozen others occupied in the same way in various parts of the quarter, for about every third Chinaman runs a lottery, and the balance of the tribe "buck" at it. "Tom," who speaks faultless English, and used to be chief and only cook to the *Territorial Enterprise*, when the establishment kept bachelor's hall two years ago, said that "Sometime Chinaman buy ticket one dollar hap, ketch um two tree hundred, sometime no ketch um anyting; lottery like one man fight um seventy — may-be he whip, may-be he get whip heself, welly good." However, the percentage being sixty-nine against him, the chances are, as a general thing, that "he get whip heself." We could not see that these lotteries differed in any respect from our own, save that the figures being Chinese, no ignorant white man might ever hope to succeed in telling "t'other from which;" the manner of drawing is similar to ours.

Mr. See Yup keeps a fancy store on Live Fox street. He sold us fans of white feathers, gorgeously ornamented; perfumery that smelled like Limburger cheese, Chinese pens, and watch-charms made of a stone unscratchable with steel instruments, yet polished and tinted like the inner coat of a sea-shell.* As tokens of his esteem, See Yup presented the party with gaudy plumes made of gold tinsel and trimmed with peacocks' feathers.

We ate chow-chow with chop-sticks in the celestial restaurants; our comrade chided the moon-eyed damsels in front of the houses for their want of feminine reserve; we received protecting Josh-lights from our hosts and "dickered" for a pagan god or two. Finally, we were impressed with the genius of a Chinese bookkeeper; he figured up his accounts on a machine like a gridiron with buttons strung on its bars; the different rows represented units, tens, hundreds, and thousands. He fingered them with incredible rapidity — in fact, he pushed them from

---

* A peculiar species of the "jade-stone" — to a Chinaman peculiarly precious.

place to place as fast as a musical professor's fingers travel over the keys of a piano.

They are a kindly-disposed, well-meaning race, and are respected and well treated by the upper classes, all over the Pacific coast. No Californian *gentleman or lady* ever abuses or oppresses a China-man, under any circumstances, an explanation that seems to be much needed in the East. Only the scum of the population do it — they and their children; they, and, naturally and consistently, the policemen and politicians, likewise, for these are the dust-licking pimps and slaves of the scum, there as well as elsewhere in America.

CHAPTER XIV.

I BEGAN to get tired of staying in one place so long. There was no longer satisfying variety in going down to Carson to report the proceedings of the legislature once a year, and horse-races and pumpkin-shows once in three months (they had got to raising pumpkins and potatoes in Washoe Valley, and of course one of the first achievements of the legislature was to institute a ten-thousand-dollar agricultural fair to show off forty dollars' worth of those pumpkins in — however, the Territorial legislature was usually spoken of as the "asylum"). I wanted to see San Francisco. I wanted to go somewhere. I wanted — I did not know *what* I wanted. I had the "spring fever" and wanted a change, principally, no doubt. Besides, a convention had framed a State Constitution; nine men out of every ten wanted an office; I believed that these gentlemen would "treat" the moneyless and the irresponsible among the population into adopting the constitution and thus well-nigh killing the country (it could not well carry such a load as a State government, since it had nothing

(136)

to tax that could stand a tax, for undeveloped mines could not, and there were not fifty developed ones in the land, there was but little realty to tax, and it did seem as if nobody was ever going to think of the simple salvation of inflicting a money penalty on murder). I believed that a State government would destroy the " flush times," and I wanted to get away. I believed that the mining stocks I had on hand would soon be worth $100,000, and thought if they reached that before the constitution was adopted, I would sell out and make myself secure from the crash the change of government was going to bring. I considered $100,000 sufficient to go home with decently, though it was but a small amount compared to what I had been expecting to return with. I felt rather downhearted about it, but I tried to comfort myself with the reflection that with such a sum I could not fall into want. About this time a schoolmate of mine, whom I had not seen since boyhood, came tramping in on foot from Reese River, a very allegory of Poverty. The son of wealthy parents, here he was, in a strange land, hungry, bootless, mantled in an ancient horse-blanket, roofed with a brimless hat, and so generally and so extravagantly dilapidated that he could have " taken the shine out of the Prodigal Son himself," as he pleasantly remarked. He wanted to borrow forty-six dollars — twenty-six to take him to San Francisco, and twenty for something else; to buy some soap with, maybe, for he needed it. I found

I had but little more than the amount wanted, in
my pocket; so I stepped in and borrowed forty-six
dollars of a banker (on twenty days' time, without
the formality of a note), and gave it him, rather
than walk half a block to the office, where I had
some specie laid up. If anybody had told me that
it would take me two years to pay back that forty-
six dollars to the banker (for I did not expect it of
the Prodigal, and was not disappointed), I would
have felt injured. And so would the banker.

I wanted a change. I wanted variety of some
kind. It came. Mr. Goodman went away for a
week and left me the post of chief editor. It destroyed me. The first day, I wrote my " leader "
in the forenoon. The second day, I had no subject
and put it off till the afternoon. The third day I
put it off till evening, and then copied an elaborate
editorial out of the " American Cyclopædia," that
steadfast friend of the editor, all over this land.
The fourth day I " fooled around " till midnight,
and then fell back on the Cyclopædia again. The
fifth day I cudgeled my brain till midnight, and then
kept the press waiting while I penned some bitter
personalities on six different people. The sixth day
I labored in anguish till far into the night and
brought forth — nothing. The paper went to press
without an editorial. The seventh day I resigned.
On the eighth, Mr. Goodman returned and found
six duels on his hands — my personalities had borne
fruit.

Nobody, except he has tried it, knows what it is to be an editor. It is easy to scribble local rubbish, with the facts all before you; it is easy to clip selections from other papers; it is easy to string out a correspondence from any locality; but it is unspeakable hardship to write editorials. *Subjects* are the trouble — the dreary lack of them, I mean. Every day, it is drag, drag, drag — think, and worry, and suffer — all the world is a dull blank, and yet the editorial columns *must* be filled. Only give the editor a *subject*, and his work is done — it is no trouble to write it up; but fancy how you would feel if you had to pump your brains dry every day in the week, fifty-two weeks in the year. It makes one low-spirited simply to think of it. The matter that each editor of a daily paper in America writes in the course of a year would fill from four to eight bulky volumes like this book! Fancy what a library an editor's work would make, after twenty or thirty years' service. Yet people often marvel that Dickens, Scott, Bulwer, Dumas, etc., have been able to produce so many books. If these authors had wrought as voluminously as newspaper editors do, the result would be something to marvel at, indeed. How editors can continue this tremendous labor, this exhausting consumption of brain fiber (for their work is creative, and not a mere mechanical laying-up of facts, like reporting), day after day and year after year, is incomprehensible. Preachers take two months' holiday in midsummer, for they find that

to produce two sermons a week is wearing, in the
long run. In truth it must be so, and is so; and
therefore, how an editor can take from ten to twenty
texts and build upon them from ten to twenty pains-
taking editorials a week and keep it up all the year
round, is farther beyond comprehension than ever.
Ever since I survived my week as editor, I have
found at least one pleasure in any newspaper that
comes to my hand; it is in admiring the long columns of editorial, and wondering to myself how in
the mischief he did it!

Mr. Goodman's return relieved me of employment, unless I chose to become a reporter again. I
could not do that; I could not serve in the ranks
after being general of the army. So I thought I
would depart and go abroad into the world somewhere. Just at this juncture, Dan, my associate in
the reportorial department, told me, casually, that
two citizens had been trying to persuade him to go
with them to New York and aid in selling a rich
silver mine which they had discovered and secured
in a new mining district in our neighborhood. He
said they offered to pay his expenses and give him
one-third of the proceeds of the sale. He had
refused to go. It was the very opportunity I
wanted. I abused him for keeping so quiet about
it, and not mentioning it sooner. He said it had
not occurred to him that I would like to go, and so
he had recommended them to apply to Marshall, the
reporter of the other paper. I asked Dan if it was

a good, honest mine, and no swindle. He said the men had shown him nine tons of the rock, which they had got out to take to New York, and he could cheerfully say that he had seen but little rock in Nevada that was richer; and, moreover, he said that they had secured a tract of valuable timber and a mill-site, near the mine. My first idea was to kill Dan. But I changed my mind, notwithstanding I was so angry, for I thought maybe the chance was not yet lost. Dan said it was by no means lost; that the men were absent at the mine again, and would not be in Virginia to leave for the East for some ten days; that they had requested him to do the talking to Marshall, and he had promised that he would either secure Marshall or somebody else for them by the time they got back; he would now say nothing to anybody till they returned, and then fulfil his promise by furnishing me to them.

It was splendid. I went to bed all on fire with excitement; for nobody had yet gone East to sell a Nevada silver mine, and the field was white for the sickle. I felt that such a mine as the one described by Dan would bring a princely sum in New York, and sell without delay or difficulty. I could not sleep, my fancy so rioted through its castles in the air. It was the " blind lead " come again.

Next day I got away, on the coach, with the usual éclat attending departures of old citizens,— for if you have only half a dozen friends out there they will make noise for a hundred rather than let you

seem to go away neglected and unregretted — and
Dan promised to keep strict watch for the men that
had the mine to sell.

The trip was signalized but by one little incident,
and that occurred just as we were about to start. A
very seedy-looking vagabond passenger got out of
the stage a moment to wait till the usual ballast of
silver bricks was thrown in. He was standing on
the pavement, when an awkward express employé,
carrying a brick weighing a hundred pounds, stumbled and let it fall on the bummer's foot. He instantly dropped on the ground and began to howl in
the most heart-breaking way. A sympathizing
crowd gathered around and were going to pull his
boot off; but he screamed louder than ever and
they desisted; then he fell to gasping, and between
the gasps ejaculated "Brandy! for Heaven's sake,
brandy!" They poured half a pint down him, and
it wonderfully restored and comforted him. Then
he begged the people to assist him to the stage,
which was done. The express people urged him to
have a doctor at their expense, but he declined, and
said that if he only had a little brandy to take along
with him, to soothe his paroxysms of pain when they
came on, he would be grateful and content. He
was quickly supplied with two bottles, and we drove
off. He was so smiling and happy after that, that
I could not refrain from asking him how he could
possibly be so comfortable with a crushed foot.

"Well," said he, "I hadn't had a drink for

twelve hours, and hadn't a cent to my name. I was most perishing — and so, when that duffer dropped that hundred-pounder on my foot, I see my chance. Got a cork leg, you know!'' and he pulled up his pantaloons and proved it.

He was as drunk as a lord all day long, and full of chucklings over his timely ingenuity.

One drunken man necessarily reminds one of another. I once heard a gentleman tell about an incident which he witnessed in a Californian bar-room. He entitled it "Ye Modest Man Taketh a Drink." It was nothing but a bit of acting, but it seemed to me a perfect rendering, and worthy of Toodles himself. The modest man, tolerably far gone with beer and other matters, enters a saloon (twenty-five cents is the price for anything and everything, and specie the only money used) and lays down a half dollar; calls for whisky and drinks it; the barkeeper makes change and lays the quarter in a wet place on the counter; the modest man fumbles at it with nerveless fingers, but it slips and the water holds it; he contemplates it, and tries again; same result; observes that people are interested in what he is at, blushes; fumbles at the quarter again — blushes — puts his forefinger carefully, slowly down, to make sure of his aim — pushes the coin toward the barkeeper, and says with a sigh:

"('ic!) Gimme a cigar!''

Naturally, another gentleman present told about

another drunken man. He said he reeled toward
home late at night; made a mistake and entered the
wrong gate; thought he saw a dog on the stoop;
and it was — an iron one. He stopped and considered; wondered if it was a dangerous dog; ven-
tured to say "Be (hic) begone!" No effect.
Then he approached warily, and adopted concilia-
tion; pursed up his lips and tried to whistle, but
failed; still approached, saying, "Poor dog!—
doggy, doggy, doggy!— poor doggy-dog!" Got
up on the stoop, still petting with fond names, till
master of the advantages; then exclaimed, "Leave,
you thief!"— planted a vindictive kick in his ribs,
and went head-over-heels overboard, of course. A
pause; a sigh or two of pain, and then a remark in
a reflective voice:

"Awful solid dog. What could he ben eating?
('ic!) Rocks, p'raps. Such animals is dangerous.
'At's what *I* say — they're dangerous. If a man —
('ic!)— if a man wants to feed a dog on rocks, let
him *feed* him on rocks; 'at's all right; but let him
keep him at *home* — not have him layin' round
promiscuous, where ('ic!) where people's liable to
stumble over him when they ain't noticin'!"

It was not without regret that I took a last look
at the tiny flag (it was thirty-five feet long and ten
feet wide) fluttering like a lady's handkerchief from
the topmost peak of Mount Davidson, two thousand
feet above Virginia's roofs, and felt that doubtless I
was bidding a permanent farewell to a city which

had afforded me the most vigorous enjoyment of
life I had ever experienced. And this reminds me
of an incident which the dullest memory Virginia
could boast at the time it happened must vividly
recall, at times, till its possessor dies. Late one
summer afternoon we had a rain shower. That was
astonishing enough, in itself, to set the whole town
buzzing, for it only rains (during a week or two
weeks) in the winter in Nevada, and even then not
enough at a time to make it worth while for any
merchant to keep umbrellas for sale. But the rain
was not the chief wonder. It only lasted five or ten
minutes; while the people were still talking about it
all the heavens gathered to themselves a dense
blackness as of midnight. All the vast eastern front
of Mount Davidson, overlooking the city, put on
such a funereal gloom that only the nearness and
solidity of the mountain made its outlines even
faintly distinguishable from the dead blackness of
the heavens they rested against. This unaccustomed sight turned all eyes toward the mountain;
and as they looked, a little tongue of rich golden
flame was seen waving and quivering in the heart of
the midnight, away up on the extreme summit! In
a few minutes the streets were packed with people,
gazing with hardly an uttered word, at the one
brilliant mote in the brooding world of darkness.
It flickered like a candle-flame, and looked no larger;
but with such a background it was wonderfully
bright, small as it was. It was the flag! — though

10

no one suspected it at first, it seemed so like a supernatural visitor of some kind — a mysterious messenger of good tidings, some were fain to believe. It was the nation's emblem transfigured by the departing rays of a sun that was entirely palled from view; and on no other object did the glory fall, in all the broad panorama of mountain ranges and deserts. Not even upon the staff of the flag — for that, a needle in the distance at any time, was now untouched by the light and undistinguishable in the gloom. For a whole hour the weird visitor winked and burned in its lofty solitude, and still the thousands of uplifted eyes watched it with fascinated interest. How the people were wrought up! The superstition grew apace that this was a mystic courier come with great news from the war — the poetry of the idea excusing and commending it — and on it spread, from heart to heart, from lip to lip, and from street to street, till there was a general impulse to have out the military and welcome the bright waif with a salvo of artillery!

And all that time one sorely-tried man, the telegraph operator, sworn to official secrecy, had to lock his lips and chain his tongue with a silence that was like to rend them; for he, and he only, of all the speculating multitude, knew the great things this sinking sun had seen that day in the East — Vicksburg fallen, and the Union arms victorious at Gettysburg!

But for the journalistic monopoly that forbade the

slightest revealment of Eastern news till a day after
its publication in the California papers, the glorified
flag on Mount Davidson would have been saluted
and re-saluted, that memorable evening, as long as
there was a charge of powder to thunder with; the
city would have been illuminated, and every man
that had any respect for himself would have got
drunk,— as was the custom of the country on all
occasions of public moment.   Even at this distant
day I cannot think of this needlessly marred supreme
opportunity without regret.   What a time we might
have had !

CHAPTER XV.

WE rumbled over the plains and valleys, climbed the Sierras to the clouds, and looked down upon summer-clad California. And I will remark here, in passing, that all scenery in California requires *distance* to give it its highest charm. The mountains are imposing in their sublimity and their majesty of form and altitude, from any point of view — but one must have distance to soften their ruggedness and enrich their tintings; a Californian forest is best at a little distance, for there is a sad poverty of variety in species, the trees being chiefly of one monotonous family — redwood, pine, spruce, fir — and so, at a near view there is a wearisome sameness of attitude in their rigid arms, stretched downward and outward in one continued and reiterated appeal to all men to " Sh!— don't say a word! — you might disturb somebody !" Close at hand, too, there is a reliefless and relentless smell of pitch and turpentine; there is a ceaseless melancholy in their sighing and complaining foliage; one walks over a soundless carpet of beaten yellow bark and dead spines of the foliage till he feels like a wandering spirit bereft of a footfall; he tires of the endless tufts of needles and yearns for substantial, shapely leaves; he looks for moss and grass to loll upon, and finds none, for where there is no bark there is naked clay and dirt, enemies to pensive musing and clean apparel. Often a grassy plain in California is what it should be, but often, too, it is best contemplated at a distance, because although its grass blades are tall, they stand up vindictively straight and self-sufficient, and are unsociably wide apart, with uncomely spots of barren sand between.

One of the queerest things I know of, is to hear tourists from " the States " go into ecstasies over the loveliness of " ever-blooming California." And they always do go into that sort of ecstasies. But perhaps they would modify them if they knew how old Californians, with the memory full upon them of the dust-covered and questionable summer greens of Californian " verdure," stand astonished, and filled with worshiping admiration, in the presence of the lavish richness, the brilliant green, the infinite freshness, the spendthrift variety of form and species and foliage that make an Eastern landscape a vision of Paradise itself. The idea of a man falling into raptures over grave and somber California, when that man has seen New England's meadow-expanses and her maples, oaks, and cathedral-windowed elms decked in summer attire, or the opaline splendors of autumn descending upon her forests, comes very near being funny — would be, in fact, but that it is so pathetic. No land with an unvarying climate can be very beautiful. The tropics are not, for all the sentiment that is wasted on them. They seem beautiful at first, but sameness impairs the charm by and by. *Change* is the handmaiden Nature requires to do her miracles with. The land that has four well-defined seasons cannot lack beauty, or pall with monotony. Each season brings a world of enjoyment and interest in the watching of its unfolding, its gradual, harmonious development, its culminating graces — and just as one begins to tire of it, it passes away and a radical change comes, with new witcheries and new glories in its train. And I think that to one in sympathy with nature, each season, in its turn, seems the loveliest.

San Francisco, a truly fascinating city to live in, is stately and handsome at a fair distance, but close at hand one notes that the architecture is mostly old-fashioned, many streets are made up of decaying, smoke-grimed, wooden houses, and the barren sand-hills toward the outskirts obtrude themselves too prominently. Even the kindly climate is sometimes pleasanter when read about than personally experienced, for a lovely, cloudless sky wears out its welcome by and by, and then when the longed-for rain does come it *stays*. Even the playful earthquake is better contemplated at a dis—

However, there are varying opinions about that.

The climate of San Francisco is mild and singularly equable. The thermometer stands at about seventy degrees the year round. It hardly changes at all. You sleep under one or two light blankets summer and winter, and never use a mosquito bar. Nobody ever wears summer clothing. You wear black broadcloth — if you have it — in August and January, just the same. It is no colder, and no warmer, in the one month than the other. You do not use overcoats and you do not use fans. It is as pleasant a climate as could well be contrived, take it all around, and is doubtless the most unvarying in the whole world. The wind blows there a good deal in the summer months, but then you can go over to Oakland, if you choose — three or four miles away — it does not blow there. It has only snowed twice in San Francisco in nineteen years, and then it only remained on the ground long enough to astonish the children, and set them to wondering what the feathery stuff was.

During eight months of the year, straight along, the skies are bright and cloudless, and never a drop of rain falls. But when the other four months come along, you will need to go and steal an umbrella. Because you will require it. Not just one day, but one hundred and twenty days in hardly varying succession. When you want to go visiting, or attend church, or the theater, you never look up at the clouds to see whether it is likely to rain or not — you look at the almanac. If it is winter, it will *rain* — and if it is summer, it *won't* rain, and you cannot help it. You never need a lightning-rod, because it never thunders and it never lightens. And after you have listened for six or eight weeks, every night, to the dismal monotony of those quiet rains, you will wish in your heart the thunder *would* leap and crash and roar along those drowsy skies once, and make everything alive — you will wish the prisoned lightnings *would* cleave the dull firmament asunder and light it with a blinding glare for *one* little instant. You would give *anything* to hear the old familiar thunder again and see the lightning strike somebody. And along in the summer, when you have suffered about four months of lustrous, pitiless sunshine, you are ready to go down on your knees and plead for rain — hail — snow — thunder and lightning — anything to break the monotony — you will take an earthquake, if you cannot do any better. And the chances are that you'll get it, too.

San Francisco is built on sand-hills, but they are prolific sand-hills. They yield a generous vegetation. All the rare flowers which people in " the States " rear with such patient care in parlor flower-pots and greenhouses, flourish luxuriantly in the open air there all the year round. Calla lilies, all sorts of geraniums, passion flowers, moss roses — I do not know the names of a tenth part of them. I only know that while New Yorkers are burdened with banks and drifts of snow, Californians are burdened with banks and drifts of flowers, if they only keep their hands off and let them grow. And I have heard that they have also that rarest and most curious of all the flowers, the beautiful *Espiritu Santo*, as the Spaniards call it — or flower of the Holy Spirit — though I thought it grew only in Central America — down on the Isthmus. In its cup is the daintiest little facsimile of a dove, as pure as snow. The Spaniards have a superstitious reverence for it. The blossom has been conveyed to the States, submerged in ether; and the bulb has been taken thither also, but every attempt to make it bloom after it arrived, has failed.

I have elsewhere spoken of the endless winter of Mono, California, and but this moment of the eternal spring of San Francisco. Now, if we travel a hundred miles in a straight line, we come to the eternal summer of Sacramento. One never sees summer-clothing or mosquitoes in San Francisco — but they can be found in Sacramento. Not always and unvaryingly, but about one hundred and forty-three months out of twelve years, perhaps. Flowers bloom there, always, the reader can easily believe — people suffer and sweat, and swear, morning, noon, and night, and wear out their stanchest energies fanning themselves. It gets hot there, but if you go down to Fort Yuma you will find it hotter. Fort Yuma is probably the hottest place on earth. The thermometer stays at one hundred and twenty in the shade there all the time — except when it varies and goes higher. It is a U. S. military post, and its occupants get so used to the terrific heat that they suffer without it. There is a tradition (attributed to John Phenix*) that a very, very wicked soldier died

---

* It has been purloined by fifty different scribblers who were too poor to invent a fancy but not ashamed to steal one. — M. T.

there, once, and of course, went straight to the
hottest corner of perdition,— and the next day he
telegraphed back for his blankets. There is no
doubt about the truth of this statement. There can
be no doubt about it. I have seen the place where
that soldier used to board. In Sacramento it is fiery
summer always, and you can gather roses, and eat
strawberries and ice-cream, and wear white linen
clothes, and pant and perspire, at eight or nine
o'clock in the morning, and then take the cars, and
at noon put on your furs and your skates, and go
skimming over frozen Donner Lake, seven thousand
feet above the valley, among snowbanks fifteen feet
deep, and in the shadow of grand mountain peaks
that lift their frosty crags ten thousand feet above
the level of the sea. There is a transition for you!
Where will you find another like it in the western
hemisphere? And some of us have swept around
snow-walled curves of the Pacific Railroad in that
vicinity, six thousand feet above the sea, and looked
down as the birds do, upon the deathless summer of
the Sacramento Valley, with its fruitful fields, its
feathery foliage, its silver streams, all slumbering in
the mellow haze of its enchanted atmosphere, and all
infinitely softened and spiritualized by distance — a
dreamy, exquisite glimpse of fairyland, made all the
more charming and striking that it was caught
through a forbidden gateway of ice and snow, and
savage crags and precipices.

CHAPTER XVI.

IT was in this Sacramento Valley, just referred to,
that a deal of the most lucrative of the early
gold mining was done, and you may still see, in
places, its grassy slopes and levels torn and guttered
and disfigured by the avaricious spoilers of fifteen
and twenty years ago. You may see such dis-
figurements far and wide over California — and in
some such places, where only meadows and forests
are visible — not a living creature, not a house, no
stick or stone or remnant of a ruin, and not a sound,
not even a whisper to disturb the Sabbath stillness —
you will find it hard to believe that there stood at
one time a fiercely-flourishing little city, of two
thousand or three thousand souls, with its newspaper, fire company, brass band, volunteer militia,
bank, hotels, noisy Fourth of July processions and
speeches, gambling hells crammed with tobacco
smoke, profanity, and rough-bearded men of all
nations and colors, with tables heaped with gold
dust sufficient for the revenues of a German princi-
pality — streets crowded and rife with business —
town lots worth four hundred dollars a front foot —

labor, laughter, music, dancing, swearing, fighting,
shooting, stabbing — a bloody inquest and a man
for breakfast every morning — *everything* that delights and adorns existence — all the appointments
and appurtenances of a thriving and prosperous and
promising young city,— and *now* nothing is left of
it all but a lifeless, homeless solitude.   The men are
gone, the houses have vanished, even the *name* of
the place is forgotten.   In no other land, in modern
times, have towns so absolutely died and disappeared, as in the old mining regions of California.

It was a driving, vigorous, restless population in
those days.   It was a *curious* population.   It was
the *only* population of the kind that the world has
ever seen gathered together, and it is not likely that
the world will ever see its like again.   For, observe,
it was an assemblage of two hundred thousand *young*
men — not simpering, dainty, kid-gloved weaklings,
but stalwart, muscular, dauntless young braves,
brimful of push and energy, and royally endowed
with every attribute that goes to make up a peerless
and magnificent manhood — the very pick and
choice of the world's glorious ones.   No women,
no children, no gray and stooping veterans,— none
but erect, bright-eyed, quick-moving, strong-handed
young giants — the strangest population, the finest
population, the most gallant host that ever trooped
down the startled solitudes of an unpeopled land.
And where are they now?   Scattered to the ends of
the earth — or prematurely aged and decrepit — or

shot or stabbed in street affrays — or dead of disappointed hopes and broken hearts — all gone, or nearly all — victims devoted upon the altar of the golden calf — the noblest holocaust that ever wafted its sacrificial incense heavenward. It is pitiful to think upon.

It was a splendid population — for all the slow, sleepy, sluggish-brained sloths staid at home — you never find that sort of people among pioneers — you cannot build pioneers out of that sort of material. It was that population that gave to California a name for getting up astounding enterprises and rushing them through with a magnificent dash and daring and a recklessness of cost or consequences, which she bears unto this day — and when she projects a new surprise, the grave world smiles as usual, and says " Well, that is California all over."

But they were rough in those times! They fairly reveled in gold, whisky, fights, and fandangoes, and were unspeakably happy. The honest miner raked from a hundred to a thousand dollars out of his claim a day, and what with the gambling dens and the other entertainments, he hadn't a cent the next morning, if he had any sort of luck. They cooked their own bacon and beans, sewed on their own buttons, washed their own shirts — blue woollen ones; and if a man wanted a fight on his hands without any annoying delay, all he had to do was to appear in public in a white shirt or a stove-pipe hat, and he would be accommodated. For those people hated aristocrats. They had a particular and malignant animosity toward what they called a "biled shirt."

It was a wild, free, disorderly, grotesque society! *Men* — only swarming hosts of stalwart *men* — nothing juvenile, nothing feminine, visible anywhere!

In those days miners would flock in crowds to catch a glimpse of that rare and blessed spectacle, a woman! Old inhabitants tell how, in a certain camp, the news went abroad early in the morning that a woman was come! They had seen a calico dress hanging out of a wagon down at the camping-ground — sign of emigrants from over the great plains. Everybody went down there, and a shout went up when an actual, *bona fide* dress was discovered fluttering in the wind! The male emigrant was visible. The miners said:

"Fetch her out!"

He said: "It is my wife, gentlemen — she is sick — we have been robbed of money, provisions, everything, by the Indians — we want to rest."

"Fetch her out! We've got to see her!"

"But, gentlemen, the poor thing, she —"

"FETCH HER OUT!"

He "fetched her out," and they swung their hats and sent up three rousing cheers and a tiger; and they crowded around and gazed at her, and touched her dress, and listened to her voice with the look of men who listened to a *memory* rather than a present reality — and then they collected twenty-five hun-

HE "FETCHED HER OUT"

dred dollars in gold and gave it to the man, and
swung their hats again and gave three more cheers,
and went home satisfied.

Once I dined in San Francisco with the family of
a pioneer, and talked with his daughter, a young
lady whose first experience in San Francisco was an
adventure, though she herself did not remember it,
as she was only two or three years old at the time.
Her father said that, after landing from the ship,
they were walking up the street, a servant leading
the party with the little girl in her arms. And
presently a huge miner, bearded, belted, spurred,
and bristling with deadly weapons — just down from
a long campaign in the mountains, evidently —
barred the way, stopped the servant, and stood
gazing, with a face all alive with gratification and
astonishment. Then he said, reverently:

"Well, if it ain't a child!" And then he
snatched a little leather sack out of his pocket and
said to the servant:

"There's a hundred and fifty dollars in dust,
there, and I'll give it to you to let me kiss the
child!"

That anecdote is *true*.

But see how things change. Sitting at that
dinner-table, listening to that anecdote, if I had
offered double the money for the privilege of kissing
the same child, I would have been refused. Seventeen added years have far more than doubled the
price.

And while upon this subject I will remark that once in Star City, in the Humboldt Mountains, I took my place in a sort of long, post-office single file of miners, to patiently await my chance to peep through a crack in the cabin and get a sight of the splendid new sensation — a genuine, live Woman! And at the end of half of an hour my turn came, and I put my eye to the crack, and there she was, with one arm akimbo, and tossing flapjacks in a frying-pan with the other. And she was one hundred and sixty-five* years old, and hadn't a tooth in her head.

---

* Being in calmer mood, now, I voluntarily knock off a hundred from that. — M. T.

CHAPTER XVII.

FOR a few months I enjoyed what to me was an entirely new phase of existence — a butterfly idleness; nothing to do, nobody to be responsible to, and untroubled with financial uneasiness. I fell in love with the most cordial and sociable city in the Union. After the sage-brush and alkali deserts of Washoe, San Francisco was Paradise to me. I lived at the best hotel, exhibited my clothes in the most conspicuous places, infested the opera, and learned to seem enraptured with music which oftener afflicted my ignorant ear than enchanted it, if I had had the vulgar honesty to confess it. However, I suppose I was not greatly worse than the most of my countrymen in that. I had longed to be a butterfly, and I was one at last. I attended private parties in sumptuous evening dress, simpered and aired my graces like a born beau, and polked and schottisched with a step peculiar to myself — and the kangaroo. In a word, I kept the due state of a man worth a hundred thousand dollars (prospectively), and likely to reach absolute affluence when that silver-mine sale should be ultimately achieved in

11.**

the East. I spent money with a free hand, and
meantime watched the stock sales with an interested
eye and looked to see what might happen in Nevada.

Something very important happened. The prop-
erty-holders of Nevada voted against the State Constitution; but the folks who had nothing to lose
were in the majority, and carried the measure over
their heads. But after all it did not immediately
look like a disaster, though unquestionably it was
one. I hesitated, calculated the chances, and then
concluded not to sell. Stocks went on rising;
speculation went mad; bankers, merchants, lawyers,
doctors, mechanics, laborers, even the very washer-
women and servant girls, were putting up their
earnings on silver stocks, and every sun that rose in
the morning went down on paupers enriched and
rich men beggared. What a gambling carnival it
was! Gould & Curry soared to six thousand
hundred dollars a foot! And then — all of a sudden, out went the bottom and everything and
everybody went to ruin and destruction! The
wreck was complete. The bubble scarcely left a
microscopic moisture behind it. I was an early
beggar and a thorough one. My hoarded stocks
were not worth the paper they were printed on. I
threw them all away. I, the cheerful idiot that had
been squandering money like water, and thought
myself beyond the reach of misfortune, had not now
as much as fifty dollars when I gathered together
my various debts and paid them. I removed from

the hotel to a very private boarding-house. I took
a reporter's berth and went to work. I was not
entirely broken in spirit, for I was building confidently on the sale of the silver mine in the East.
But I could not hear from Dan. My letters mis-
carried or were not answered:

One day I did not feel vigorous and remained
away from the office. The next day I went down
toward noon as usual, and found a note on my desk
which had been there twenty-four hours. It was
signed "Marshall" — the Virginia reporter — and
contained a request that I should call at the hotel
and see him and a friend or two that night, as they
would sail for the East in the morning. A post-
script added that their errand was a big mining
speculation! I was hardly ever so sick in my life.
I abused myself for leaving Virginia and entrusting
to another man a matter I ought to have attended to
myself; I abused myself for remaining away from
the office on the one day of all the year that I should
have been there. And thus berating myself I trotted
a mile to the steamer wharf and arrived just in time
to be too late. The ship was in the stream and
under way.

I comforted myself with the thought that may be
the speculation would amount to nothing — poor
comfort at best — and then went back to my slavery,
resolved to put up with my thirty-five dollars a week
and forget all about it.

A month afterward I enjoyed my first earthquake.

K**

It was one which was long called the "great" earthquake, and is doubtless so distinguished till this day. It was just after noon, on a bright October day. I was coming down Third street. The only objects in motion anywhere in sight in that thickly-built and populous quarter, were a man in a buggy behind me, and a street car wending slowly up the cross street. Otherwise, all was solitude and a Sabbath stillness. As I turned the corner, around a frame house, there was a great rattle and jar, and it occurred to me that here was an item!— no doubt a fight in that house. Before I could turn and seek the door, there came a really terrific shock; the ground seemed to roll under me in waves, inter-rupted by a violent joggling up and down, and there was a heavy grinding noise as of brick houses rub-bing together. I fell up against the frame house and hurt my elbow. I knew what it was, now, and from mere reportorial instinct, nothing else, took out my watch and noted the time of day; at that moment a third and still severer shock came, and as I reeled about on the pavement trying to keep my footing, I saw a sight! The entire front of a tall four-story brick building in Third street sprung out-ward like a door and fell sprawling across the street, raising a dust like a great volume of smoke! And here came the buggy — overboard went the man, and in less time than I can tell it the vehicle was distributed in small fragments along three hundred yards of street. One could have fancied that somebody had fired a charge of chair-rounds and rags down the thoroughfare. The street car had stopped, the horses were rearing and plunging, the passengers were pouring out at both ends, and one fat man had crashed half way through a glass window on one side of the car, got wedged fast and was squirming and screaming like an impaled madman. Every door of every house, as far as the eye could reach, was vomiting a stream of human beings; and almost before one could execute a wink and begin another, there was a massed multitude of people stretching in endless procession down every street my position commanded. Never was solemn solitude turned into teeming life quicker.

Of the wonders wrought by "the great earthquake," these were all that came under my eye; but the tricks it did, elsewhere, and far and wide over the town, made toothsome gossip for nine days. The destruction of property was trifling — the injury to it was widespread and somewhat serious.

The "curiosities" of the earthquake were simply endless. Gentlemen and ladies who were sick, or were taking a siesta, or had dissipated till a late hour and were making up lost sleep, thronged into the public streets in all sorts of queer apparel, and some without any at all. One woman who had been washing a naked child, ran down the street holding it by the ankles as if it were a dressed turkey. Prominent citizens who were supposed to keep the Sabbath strictly, rushed out of saloons in their shirt-

sleeves, with billiard cues in their hands. Dozens
of men with necks swathed in napkins rushed from
barber-shops, lathered to the eyes or with one cheek
clean shaved and the other still bearing a hairy
stubble. Horses broke from stables, and a frightened dog rushed up a short attic ladder and out on
to a roof, and when his scare was over had not the
nerve to go down again the same way he had gone
up. A prominent editor flew downstairs, in the
principal hotel, with nothing on but one brief under-
garment — met a chambermaid, and exclaimed:

" Oh, what *shall* I do! Where shall I go!"

She responded with naïve serenity:

" If you have no choice, you might try a clothing
store!"

A certain foreign consul's lady was the acknowl-
edged leader of fashion, and every time she appeared
in anything new or extraordinary, the ladies in the
vicinity made a raid on their husbands' purses and
arrayed themselves similarly. One man, who had
suffered considerably and growled accordingly, was
standing at the window when the shocks came, and
the next instant the consul's wife, just out of the
bath, fled by with no other apology for clothing
than — a bath-towel! The sufferer rose superior to
the terrors of the earthquake, and said to his wife:

" Now *that* is something *like!* Get out your
towel, my dear!"

The plastering that fell from ceilings in San Francisco that day, would have covered several acres of

ground. For some days afterward, groups of eye-
ing and pointing men stood about many a building,
looking at long zigzag cracks that extended from
the eaves to the ground. Four feet of the tops of
three chimneys on one house were broken square off
and turned around in such a way as to completely stop
the draft. A crack a hundred feet long gaped open
six inches wide in the middle of one street and then
shut together again with such force as to ridge up the
meeting earth like a slender grave. A lady, sitting
in her rocking and quaking parlor, saw the wall part
at the ceiling, open and shut twice, like a mouth,
and then drop the end of a brick on the floor like a
tooth. She was a woman easily disgusted with
foolishness, and she arose and went out of there.
One lady who was coming down stairs was astonished to see a bronze Hercules lean forward on its
pedestal as if to strike her with its club. They both
reached the bottom of the flight at the same time,—
the woman insensible from the fright. Her child,
born some little time afterward, was club-footed.
However — on second thought,— if the reader sees
any coincidence in this, he must do it at his own risk.

The first shock brought down two or three huge
organ-pipes in one of the churches. The minister,
with uplifted hands, was just closing the services
He glanced up, hesitated, and said:

"However, we will omit the benediction!"— and
the next instant there was a vacancy in the atmosphere where he had stood.

After the first shock, an Oakland minister said:

"Keep your seats! There is no better place to die than this"—

And added, after the third:

"But outside is good enough!" He then skipped out at the back door.

Such another destruction of mantel ornaments and toilet bottles as the earthquake created, San Francisco never saw before. There was hardly a girl or a matron in the city but suffered losses of this kind. Suspended pictures were thrown down, but oftener still, by a curious freak of the earthquake's humor, they were whirled completely around with their faces to the wall! There was great difference of opinion, at first, as to the course or direction the earthquake traveled, but water that splashed out of various tanks and buckets settled that. Thousands of people were made so seasick by the rolling and pitching of floors and streets that they were weak and bedridden for hours, and some few for even days afterward. Hardly an individual escaped nausea entirely.

The queer earthquake-episodes that formed the staple of San Francisco gossip for the next week would fill a much larger book than this, and so I will diverge from the subject.

By and by, in the due course of things, I picked up a copy of the *Enterprise* one day, and fell under this cruel blow:

NEVADA MINES IN NEW YORK.—G. M. Marshall, Sheba Hurst, and

Amos H. Rose, who left San Francisco last July for New York city, with ores from mines in Pine Wood District, Humboldt County, and on the Reese River range, have disposed of a mine containing six thousand feet and called the Pine Mountains Consolidated, for the sum of $3,000,-000. The stamps on the deed, which is now on its way to Humboldt County, from New York, for record, amounted to $3,000, which is said to be the largest amount of stamps ever placed on one document. A working capital of $1,000,000 has been paid into the treasury, and machinery has already been purchased for a large quartz mill, which will be put up as soon as possible. The stock in this company is all full paid and entirely unassessable. The ores of the mines in this district somewhat resemble those of the Sheba mine in Humboldt. Sheba Hurst, the discoverer of the mines, with his friends corralled all the best leads and all the land and timber they desired before making public their whereabouts. Ores from there, assayed in this city, showed them to be exceedingly rich in silver and gold — silver predominating. There is an abundance of wood and water in the District. We are glad to know that New York capital has been enlisted in the development of the mines of this region. Having seen the ores and assays, we are satisfied that the mines of the District are very valuable — anything but wildcat.

Once more native imbecility had carried the day, and I had lost a million! It was the "blind lead" over again.

Let us not dwell on this miserable matter. If I were inventing these things, I could be wonderfully humorous over them; but they are too true to be talked of with hearty levity, even at this distant day.* Suffice it that I so lost heart, and so yielded

---

* True, and yet not exactly as given in the above figures, possibly. I saw Marshall, months afterward, and although he had plenty of money he did not claim to have captured an entire *million*. In fact I gathered that he had not then received $50,000. Beyond that figure his fortune appeared to consist of uncertain vast expectations rather than prodigious certainties. However, when the above item appeared in

myself up to repinings and sighings and foolish regrets, that I neglected my duties and became about
worthless, as a reporter for a brisk newspaper. And
at last one of the proprietors took me aside, with a
charity I still remember with considerable respect,
and gave me an opportunity to resign my berth and
so save myself the disgrace of a dismissal.

---

print I put full faith in it, and incontinently wilted and went to seed
under it.

CHAPTER XVIII.

FOR a time I wrote literary screeds for the *Golden Era*. C. H. Webb had established a very excellent literary weekly called the *Californian*, but high merit was no guaranty of success; it languished, and he sold out to three printers, and Bret Harte became editor at $20 a week, and I was employed to contribute an article a week at $12. But the journal still languished, and the printers sold out to Captain Ogden, a rich man and a pleasant gentleman who chose to amuse himself with such an expensive luxury without much caring about the cost of it. When he grew tired of the novelty, he re-sold to the printers, the paper presently died a peaceful death, and I was out of work again. I would not mention these things but for the fact that they so aptly illustrate the ups and downs that characterize life on the Pacific coast. A man could hardly stumble into such a variety of queer vicissitudes in any other country.

For two months my sole occupation was avoiding acquaintances; for during that time I did not earn a penny, or buy an article of any kind, or pay my

board. I became a very adept at "slinking." I
slunk from back street to back street, I slunk away
from approaching faces that looked familiar, I slunk
to my meals, ate them humbly and with a mute
apology for every mouthful I robbed my generous
landlady of, and at midnight, after wanderings that
were but slinkings away from cheerfulness and light,
I slunk to my bed. I felt meaner, and lowlier, and
more despicable than the worms. During all this
time I had but one piece of money — a silver ten-
cent-piece — and I held to it and would not spend it
on any account, lest the consciousness coming strong
upon me that I was *entirely* penniless, might suggest
suicide. I had pawned everything but the clothes I
had on; so I clung to my dime desperately, till it
was smooth with handling.

However, I am forgetting. I did have one other
occupation beside that of " slinking." It was the
entertaining of a collector (and being entertained by
him), who had in his hands the Virginia banker's
bill for the forty-six dollars which I had loaned my
schoolmate, the " Prodigal." This man used to
call regularly once a week and dun me, and sometimes oftener. He did it from sheer force of habit,
for he knew he could get nothing. He would get
out his bill, calculate the interest for me, at five per
cent. a month, and show me clearly that there was
no attempt at fraud in it and no mistakes; and then
plead, and argue and dun with all his might for any
sum — any little trifle — even a dollar — even half a

dollar, on account. Then his duty was accomplished and his conscience free. He immediately dropped the subject there always; got out a couple of cigars and divided, put his feet in the window, and then we would have a long, luxurious talk about everything and everybody, and he would furnish me a world of curious dunning adventures out of the ample store in his memory. By and by he would clap his hat on his head, shake hands and say briskly:

"Well, business is business — can't stay with you always!"— and was off in a second.

The idea of pining for a dun! And yet I used to long for him to come, and would get as uneasy as any mother if the day went by without his visit, when I was expecting him. But he never collected that bill, at last, nor any part of it. I lived to pay it to the banker myself.

Misery loves company. Now and then at night, in out-of-the-way, dimly-lighted places, I found myself happening on another child of misfortune. He looked so seedy and forlorn, so homeless and friendless and forsaken, that I yearned toward him as a brother. I wanted to claim kinship with him and go about and enjoy our wretchedness together. The drawing toward each other must have been mutual; at any rate we got to falling together oftener, though still seemingly by accident; and although we did not speak or evince any recognition, I think the dull anxiety passed out of both of us when we saw each other, and then for several hours we would idle along contentedly, wide apart, and glancing furtively in at home lights and fireside gatherings, out of the night shadows, and very much enjoying our dumb companionship.

Finally we spoke, and were inseparable after that. For our woes were identical, almost. He had been a reporter too, and lost his berth, and this was his experience, as nearly as I can recollect it. After losing his berth, he had gone down, down, down, with never a halt; from a boarding-house on Russian Hill to a boarding-house in Kearney street; from thence to Dupont; from thence to a low sailor den; and from thence to lodgings in goods boxes and empty hogsheads near the wharves. Then, for a while, he had gained a meager living by sewing up bursted sacks of grain on the piers; when that failed he had found food here and there as chance threw it in his way. He had ceased to show his face in daylight, now, for a reporter knows everybody, rich and poor, high and low, and cannot well avoid familiar faces in the broad light of day.

This mendicant Blucher — I call him that for convenience — was a splendid creature. He was full of hope, pluck, and philosophy; he was well read and a man of cultivated taste; he had a bright wit and was a master of satire; his kindliness and his generous spirit made him royal in my eyes and changed his curbstone seat to a throne and his damaged hat to a crown.

He had an adventure once, which sticks fast in
my memory as the most pleasantly grotesque that
ever touched my sympathies. He had been without
a penny for two months. He had shirked about
obscure streets, among friendly dim lights, till the
thing had become second nature to him. But at
last he was driven abroad in daylight. The cause
was sufficient; *he had not tasted food for forty-eight
hours*, and he could not endure the misery of his
hunger in idle hiding. He came along a back street,
glowering at the loaves in bake-shop windows, and
feeling that he could trade his life away for a morsel
to eat. The sight of the bread doubled his hunger;
but it was good to look at it, anyhow, and imagine
what one might do if one only had it. Presently,
in the middle of the street, he saw a shining spot —
looked again — did not, and could not, believe his
eyes — turned away, to try them, then looked again.
It was a verity — no vain, hunger-inspired delusion —
it was a silver dime! He snatched it — gloated over
it; doubted it — bit it — found it genuine — choked
his heart down, and smothered a halleluiah. Then he
looked around — saw that nobody was looking at him
— threw the dime down where it was before — walked
away a few steps, and approached again, pretending
he did not know it was there, so that he could re-enjoy
the luxury of finding it. He walked around it, view-
ing it from different points; then sauntered about
with his hands in his pockets, looking up at the signs
and now and then glancing at it and feeling the old

12**

thrill again. Finally he took it up, and went away, fondling it in his pocket. He idled through unfrequented streets, stopping in doorways and corners to take it out and look at it. By and by, he went home to his lodgings — an empty queensware hogshead,— and employed himself till night trying to make up his mind what to buy with it. But it was hard to do. To get the most for it was the idea. He knew that at the Miner's Restaurant he could get a plate of beans and a piece of bread for ten cents; or a fish-ball and some few trifles, but they gave " no bread with one fish-ball " there. At French Pete's he could get a veal cutlet, plain, and some radishes and bread, for ten cents; or a cup of coffee — a pint at least — and a slice of bread; but the slice was not thick enough by the eighth of an inch, and sometimes they were still more criminal than that in the cutting of it. At seven o'clock his hunger was wolfish; and still his mind was not made up. He turned out and went up Merchant street, still ciphering; and chewing a bit of stick, as is the way of starving men. He passed before the lights of Martin's restaurant, the most aristocratic in the city, and stopped. It was a place where he had often dined, in better days, and Martin knew him well. Standing aside, just out of the range of the light, he worshiped the quails and steaks in the show window, and imagined that maybe the fairy times were not gone yet and some prince in disguise would come along presently and tell him to go in there and take whatever he wanted.   He chewed his stick with a hungry interest as he warmed to his subject.   Just at this juncture he was conscious of some one at his side, sure enough; and then a finger touched his arm.   He looked up, over his shoulder, and saw an apparition — a very allegory of Hunger! It was a man six feet high, gaunt, unshaven, hung with rags; with a haggard face and sunken cheeks, and eyes that pleaded piteously.   This phantom said:

"Come with me — please."

He locked his arm in Blucher's and walked up the street to where the passengers were few and the light not strong, and then facing about, put out his hands in a beseeching way, and said:

"Friend — stranger — look at me!   Life is easy to you — you go about, placid and content, as I did once, in my day — you have been in there, and eaten your sumptuous supper, and picked your teeth, and hummed your tune, and thought your pleasant thoughts, and said to yourself it is a good world — but you've never *suffered!*   You don't know what trouble is — you don't know what misery is — nor hunger!   Look at me!   Stranger, have pity on a poor, friendless, homeless dog!   As God is my judge, I have not tasted food for eight and forty hours! — look in my eyes and see if I lie! Give me the least trifle in the world to keep me from starving — anything — twenty-five cents!   Do it, stranger — do it, *please.*   It will be nothing to

12

you, but life to me. Do it, and I will go down on my knees and lick the dust before you! I will kiss your footprints — I will worship the very ground you walk on! Only twenty-five cents! I am famishing — perishing — starving by inches! For God's sake don't desert me!"

Blucher was bewildered — and touched, too — stirred to the depths. He reflected. Thought again. Then an idea struck him, and he said:

"Come with me."

He took the outcast's arm, walked him down to Martin's restaurant, seated him at a marble table, placed the bill of fare before him, and said:

"Order what you want, friend. Charge it to me, Mr. Martin."

"All right, Mr. Blucher," said Martin.

Then Blucher stepped back and leaned against the counter and watched the man stow away cargo after cargo of buckwheat cakes at seventy-five cents a plate; cup after cup of coffee, and porter-house steaks worth two dollars apiece; and when six dollars and a half's worth of destruction had been accomplished, and the stranger's hunger appeased, Blucher went down to French Pete's, bought a veal cutlet plain, a slice of bread, and three radishes, with his dime, and set to and feasted like a king!

Take the episode all around, it was as odd as any that can be culled from the myriad curiosities of Californian life, perhaps.

CHAPTER XIX.

BY and by, an old friend of mine, a miner, came down from one of the decayed mining camps of Tuolumne, California, and I went back with him. We lived in a small cabin on a verdant hillside, and there were not five other cabins in view over the wide expanse of hill and forest. Yet a flourishing city of two or three thousand population had occupied this grassy dead solitude during the flush times of twelve or fifteen years before, and where our cabin stood had once been the heart of the teeming hive, the center of the city. When the mines gave out the town fell into decay, and in a few years wholly disappeared — streets, dwellings, shops, everything — and left no sign. The grassy slopes were as green and smooth and desolate of life as if they had never been disturbed. The mere handful of miners still remaining had seen the town spring up, spread, grow, and flourish in its pride; and they had seen it sicken and die, and pass away like a dream. With it their hopes had died, and their zest of life. They had long ago resigned themselves to their exile, and ceased to correspond with their distant friends or turn longing eyes toward their early homes. They had accepted banishment, forgotten the world and been forgotten of the world. They were far from telegraphs and railroads, and they stood, as it were, in a living grave, dead to the events that stirred the globe's great populations, dead to the common interests of men, isolated and outcast from brotherhood with their kind. It was the most singular, and almost the most touching and melancholy, exile that fancy can imagine. One of my associates in this locality, for two or three months, was a man who had had a university education; but now for eighteen years he had decayed there by inches, a bearded, rough-clad, clay-stained miner, and at times, among his sighings and soliloquizings, he unconsciously interjected vaguely remembered Latin and Greek sentences — dead and musty tongues, meet vehicles for the thoughts of one whose dreams were all of the past, whose life was a failure; a tired man, burdened with the present, and indifferent to the future; a man without ties, hopes, interests, waiting for rest and the end.

In that one little corner of California is found a species of mining which is seldom or never mentioned in print. It is called "pocket mining," and I am not aware that any of it is done outside of that little corner. The gold is not evenly distributed through the surface dirt, as in ordinary placer mines, but is collected in little spots, and they are very wide apart and exceedingly hard to find, but when

you do find one you reap a rich and sudden
harvest. There are not now more than twenty
pocket-miners in that entire little region. I think I
know every one of them personally. I have known
one of them to hunt patiently about the hillsides
every day for eight months without finding gold
enough to make a snuff-box — his grocery bill running up relentlessly all the time — and then find a
pocket and take out of it two thousand dollars in
two dips of his shovel. I have known him to take
out three thousand dollars in two hours, and go and
pay up every cent of his indebtedness, then enter on
a dazzling spree that finished the last of his treasure
before the night was gone. And the next day he
bought his groceries on credit as usual, and shouldered his pan and shovel and went off to the hills
hunting pockets again happy and content. This is
the most fascinating of all the different kinds of
mining, and furnishes a very handsome percentage
of victims to the lunatic asylum.

Pocket hunting is an ingenious process. You
take a spadeful of earth from the hillside and put it
in a large tin pan and dissolve and wash it gradually
away till nothing is left but a teaspoonful of fine
sediment. Whatever gold was in that earth has
remained, because, being the heaviest, it has sought
the bottom. Among the sediment you will find
half a dozen yellow particles no larger than pin-
heads. You are delighted. You move off to one
side and wash another pan. If you find gold again,

you move to one side further, and wash a third pan.
If you find *no* gold this time, you are delighted
again, because you know you are on the right scent.
You lay an imaginary plan, shaped like a fan, with
its handle up the hill — for just where the end of
the handle is, you argue that the rich deposit lies
hidden, whose vagrant grains of gold have escaped
and been washed down the hill, spreading farther
and farther apart as they wandered. And so you
proceed up the hill, washing the earth and narrowing your lines every time the absence of gold in the
pan shows that you are outside the spread of the
fan; and at last, twenty yards up the hill your lines
have converged to a point — a single foot from that
point you cannot find any gold. Your breath comes
short and quick, you are feverish with excitement;
the dinner-bell may ring its clapper off, you pay no
attention; friends may die, weddings transpire,
houses burn down, they are nothing to you; you
sweat and dig and delve with a frantic interest — and
all at once you strike it! Up comes a spadeful of
earth and quartz that is all lovely with soiled lumps
and leaves and sprays of gold. Sometimes that
one spadeful is all — $500. Sometimes the nest
contains $10,000, and it takes you three or four
days to get it all out. The pocket-miners tell of
one nest that yielded $60,000 and two men ex-
hausted it in two weeks, and then sold the ground
for $10,000 to a party who never got $300 out of
it afterward.

The hogs are good pocket hunters. All the summer they root around the bushes, and turn up a thousand little piles of dirt, and then the miners long for the rains; for the rains beat upon these little piles and wash them down and expose the gold, possibly right over a pocket. Two pockets were found in this way by the same man in one day. One had $5,000 in it and the other $8,000. That man could appreciate it, for he hadn't had a cent for about a year.

In Tuolumne lived two miners who used to go to the neighboring village in the afternoon and return every night with household supplies. Part of the distance they traversed a trail, and nearly always sat down to rest on a great boulder that lay beside the path. In the course of thirteen years they had worn that boulder tolerably smooth, sitting on it. By and by two vagrant Mexicans came along and occupied the seat. They began to amuse themselves by chipping off flakes from the boulder with a sledge-hammer. They examined one of these flakes and found it rich with gold. That boulder paid them $800 afterward. But the aggravating circumstance was that these "Greasers" knew that there must be more gold where that boulder came from, and so they went panning up the hill and found what was probably the richest pocket that region has yet produced. It took three months to exhaust it, and it yielded $120,000. The two American miners who used to sit on the boulder

are poor yet, and they take turn about in getting up
early in the morning to curse those Mexicans — and
when it comes down to pure ornamental cursing,
the native American is gifted above the sons of men.

I have dwelt at some length upon this matter of
pocket-mining because it is a subject that is seldom
referred to in print, and therefore I judged that it
would have for the reader that interest which
naturally attaches to novelty.

CHAPTER XX.

ONE of my comrades there — another of those victims of eighteen years of unrequited toil and blighted hopes — was one of the gentlest spirits that ever bore its patient cross in a weary exile: grave and simple Dick Baker, pocket-miner of Dead-Horse Gulch. He was forty-six, gray as a rat, earnest, thoughtful, slenderly educated, slouchily dressed, and clay-soiled, but his heart was finer metal than any gold his shovel ever brought to light — than any, indeed, that ever was mined or minted.

Whenever he was out of luck and a little down-hearted, he would fall to mourning over the loss of a wonderful cat he used to own (for where women and children are not, men of kindly impulses take up with pets, for they must love something). And he always spoke of the strange sagacity of that cat with the air of a man who believed in his secret heart that there was something human about it — maybe even supernatural.

I heard him talking about this animal once. He said:

" Gentlemen, I used to have a cat here, by the

name of Tom Quartz, which you'd a took an interest in I reckon — most anybody would. I had him
here eight year — and he was the remarkablest cat
I ever see. He was a large gray one of the Tom
specie, an' he had more hard, natchral sense than
any man in this camp —'n' a *power* of dignity — he
wouldn't let the Gov'ner of Californy be familiar
with him. He never ketched a rat in his life —
'peared to be above it. He never cared for nothing
but mining. He knowed more about mining, that
cat did, than any man *I* ever, ever see. You
couldn't tell *him* noth'n' 'bout placer diggin's —'n'
as for pocket-mining, why he was just born for it.
He would dig out after me an' Jim when we went
over the hills prospect'n', and he would trot along
behind us for as much as five mile, if we went so
fur. An' he had the best judgment about mining
ground — why you never see anything like it.
When we went to work, he'd scatter a glance
around, 'n' if he didn't think much of the indications, he would give a look as much as to say,
' Well, I'll have to get you to excuse *me*,' 'n' without another word he'd hyste his nose into the air
'n' shove for home. But if the ground suited him,
he would lay low 'n' keep dark till the first pan was
washed, 'n' then he would sidle up 'n' take a look,
an' if there was about six or seven grains of gold *he*
was satisfied — he didn't want no better prospect
'n' that —'n' then he would lay down on our coats
an' snore like a steamboat till we'd struck the

pocket, an' then get up 'n' superintend. He was nearly lightnin' on superintending.

"Well, by an' by, up comes this yer quartz excitement. Everybody was into it — everybody was pick'n' 'n' blast'n' instead of shovelin' dirt on the hillside — everybody was put'n' down a shaft instead of scrapin' the surface. Noth'n' would do Jim, but *we* must tackle the ledges, too, 'n' so we did. We commenced put'n' down a shaft, 'n' Tom Quartz he begin to wonder what in the Dickens it was all about. *He* hadn't ever seen any mining like that before, 'n' he was all upset, as you may say — he couldn't come to a right understanding of it no way — it was too many for *him*. He was down on it, too, you bet you — he was down on it powerful —'n' always appeared to consider it the cussedest foolishness out. But that cat, you know, was *always* agin new-fangled arrangements — somehow he never could abide 'em. *You* know how it is with old habits. But by an' by Tom Quartz begin to git sort of reconciled a little, though he never *could* altogether understand that eternal sinkin' of a shaft an' never pannin' out anything. At last he got to comin' down in the shaft, hisself, to try to cipher it out. An' when he'd git the blues, 'n' feel kind o' scruffy, 'n' aggravated 'n' disgusted — knowin' as he did, that the bills was runnin' up all the time an' we warn't makin' a cent — he would curl up on a gunny sack in the corner an' go to sleep. Well, one day when the shaft was down about eight foot, the rock got so hard that we had to put in a blast — the first blast'n' we'd ever done since Tom Quartz was born. An' then we lit the fuse 'n' clumb out 'n' got off 'bout fifty yards — 'n' forgot 'n' left Tom Quartz sound asleep on the gunny sack. In 'bout a minute we seen a puff of smoke bust up out of the hole, 'n' then everything let go with an awful crash, 'n' about four million ton of rocks 'n' dirt 'n' smoke 'n' splinters shot up 'bout a mile an' a half into the air, an' by George, right in the dead center of it was old Tom Quartz a goin' end over end, an' a snortin' an' a sneez'n', an' a clawin' an' a reachin' for things like all possessed. But it warn't no use, you know, it warn't no use. An' that was the last we see of *him* for about two minutes 'n' a half, an' then all of a sudden it begin to rain rocks and rubbage, an' directly he come down ker-whop about ten foot off f'm where we stood. Well, I reckon he was p'raps the orneriest lookin' beast you ever see. One ear was sot back on his neck, 'n' his tail was stove up, 'n' his eye-winkers was swinged off, 'n' he was all blacked up with powder an' smoke, an' all sloppy with mud 'n' slush f'm one end to the other. Well, sir, it warn't no use to try to apologize — we couldn't say a word. He took a sort of a disgusted look at hisself, 'n' then he looked at us — an' it was just exactly the same as if he had said — ' Gents, may be *you* think it's smart to take advantage of a cat that 'ain't had no experience of quartz minin', but *I* think *different*' — an' then he turned on his heel 'n' marched off home without ever saying another word.

"That was jest his style. An' may be you won't believe it, but after that you never see a cat so pre-judiced agin quartz mining as what he was. An' by an' by when he *did* get to goin' down in the shaft agin, you'd 'a been astonished at his sagacity. The minute we'd tetch off a blast 'n' the fuse'd begin to sizzle, he'd give a look as much as to say: 'Well, I'll have to git you to excuse *me*,' an' it was sur-pris'n' the way he'd shin out of that hole 'n' go f'r a tree. Sagacity? It ain't no name for it. 'Twas *inspiration!*"

I said, "Well, Mr. Baker, his prejudice against quartz-mining *was* remarkable, considering how he came by it. Couldn't you ever cure him of it?"

"*Cure him!* No! When Tom Quartz was sot once, he was *always* sot — and you might a blowed him up as much as three million times 'n' you'd never a broken him of his cussed prejudice agin quartz mining."

The affection and the pride that lit up Baker's face when he delivered this tribute to the firmness of his humble friend of other days, will always be a vivid memory with me.

At the end of two months we had never "struck" a pocket. We had panned up and down the hillsides till they looked plowed like a field; we could have put in a crop of grain, then, but there would have been no way to get it to market. We got many

good " prospects," but when the gold gave out in
the pan and we dug down, hoping and longing, we
found only emptiness — the pocket that should have
been there was as barren as our own. At last we
shouldered our pans and shovels and struck out over
the hills to try new localities. We prospected
around Angel's Camp, in Calaveras County, during
three weeks, but had no success. Then we wandered on foot among the mountains, sleeping under
the trees at night, for the weather was mild, but still
we remained as centless as the last rose of summer.
That is a poor joke, but it is in pathetic harmony
with the circumstances, since we were so poor ourselves. In accordance with the custom of the
country, our door had always stood open and our
board welcome to tramping miners — they drifted
along nearly every day, dumped their paust shovels
by the threshold and took " pot luck " with us —
and now on our own tramp we never found cold
hospitality.

Our wanderings were wide and in many directions; and now I could give the reader a vivid description of the big trees and the marvels of the Yo
Semite — but what has this reader done to me that
I should persecute him? I will deliver him into the
hands of less conscientious tourists and take his
blessing. Let me be charitable, though I fail in all
virtues else.

Some of the phrases in the above are mining technicalities, purely,
and may be a little obscure to the general reader. In "*placer diggings*"

the gold is scattered all through the surface dirt; in "*pocket*" diggings it is concentrated in one little spot; in "*quartz*" the gold is in a solid, continuous vein of rock, enclosed between distinct walls of some other kind of stone — and this is the most laborious and expensive of all the different kinds of mining.   "*Prospecting*" is hunting for a "*placer*"; "*indications*" are signs of its presence; "*panning out*" refers to the washing process by which the grains of gold are separated from the dirt; a "*prospect*" is what one finds in the first panful of dirt — and its value determines whether it is a good or a bad prospect, and whether it is worth while to tarry there or seek further.

13**

CHAPTER XXI.

AFTER a three-months' absence, I found myself in San Francisco again, without a cent. When my credit was about exhausted (for I had become too mean and lazy, now, to work on a morning paper, and there were no vacancies on the evening journals), I was created San Francisco correspondent of the *Enterprise*, and at the end of five months I was out of debt, but my interest in my work was gone; for, my correspondence being a daily one, without rest or respite, I got unspeakably tired of it. I wanted another change. The vagabond instinct was strong upon me. Fortune favored, and I got a new berth and a delightful one. It was to go down to the Sandwich Islands and write some letters for the Sacramento *Union*, an excellent journal and liberal with employés.

We sailed in the propeller *Ajax*, in the middle of winter. The almanac called it winter, distinctly enough, but the weather was a compromise between spring and summer. Six days out of port, it became summer altogether. We had some thirty passengers; among them a cheerful soul by the name

cf Williams, and three sea-worn old whaleship captains going down to join their vessels. These latter
played euchre in the smoking-room day and night,
drank astonishing quantities of raw whisky without
being in the least affected by it, and were the hap-
piest people I think I ever saw. And then there
was " the old Admiral " — a retired whaleman. He
was a roaring, terrific combination of wind and
lightning and thunder, and earnest, whole-souled
profanity. But nevertheless he was tender-hearted
as a girl. He was a raving, deafening, devastating
typhoon, laying waste the cowering seas, but with
an unvexed refuge in the center where all comers
were safe and at rest. Nobody could know the
" Admiral " without liking him; and in a sudden
and dire emergency I think no friend of his would
know which to choose — to be cursed by him or
prayed for by a less efficient person.

His title of " Admiral " was more strictly " official " than any ever worn by a naval officer before
or since, perhaps — for it was the voluntary offering
of a whole nation, and came direct from the *people*
themselves without any intermediate red tape — the
people of the Sandwich Islands. It was a title that
came to him freighted with affection, and honor,
and appreciation of his unpretending merit. And
in testimony of the genuineness of the title it was
publicly ordained that an exclusive flag should be
devised for him and used solely to welcome his
coming and wave him God-speed in his going.

13**

From that time forth, whenever his ship was sig-
naled in the offing, or he catted his anchor and
stood out to sea, that ensign streamed from the
royal halliards on the parliament house, and the
nation lifted their hats to it with spontaneous accord.

Yet he had never fired a gun or fought a battle in
his life. When I knew him on board the *Ajax*, he
was seventy-two years old and had plowed the salt
water sixty-one of them. For sixteen years he had
gone in and out of the harbor of Honolulu in command of a whaleship, and for sixteen more had been
captain of a San Francisco and Sandwich Island
passenger packet and had never had an accident or
lost a vessel. The simple natives knew him for a
friend who never failed them, and regarded him as
children regard a father. It was a dangerous thing
to oppress them when the roaring Admiral was
around.

Two years before I knew the Admiral, he had retired from the sea on a competence, and had sworn
a colossal nine-jointed oath that he would " never go
within *smelling* distance of the salt water again as
long as he lived." And he had conscientiously
kept it. That is to say, *he* considered he had kept
it, and it would have been more than dangerous to
suggest to him, even in the gentlest way, that
making eleven long sea voyages, as a passenger,
during the two years that had transpired since he
" retired," was only keeping the general spirit of it
and not the strict letter.

The Admiral knew only one narrow line of conduct to pursue in any and all cases where there was a fight, and that was to shoulder his way straight in without an inquiry as to the rights or the merits of it, and take the part of the weaker side. And this was the reason why he was always sure to be present at the trial of any universally execrated criminal to oppress and intimidate the jury with a vindictive pantomime of what he would do to them if he ever caught them out of the box. And this was why harried cats and outlawed dogs that knew him confidently took sanctuary under his chair in time of trouble. In the beginning he was the most frantic and bloodthirsty Union man that drew breath in the shadow of the flag; but the instant the Southerners began to go down before the sweep of the Northern armies, he ran up the Confederate colors, and from that time till the end was a rampant and inexorable secessionist.

He hated intemperance with a more uncompromising animosity than any individual I have ever met, of either sex; and he was never tired of storming against it and beseeching friends and strangers alike to be wary and drink with moderation. And yet if any creature had been guileless enough to intimate that his absorbing nine gallons of " straight " whisky during our voyage was any fraction short of rigid or inflexible abstemiousness, in that self-same moment the old man would have spun him to the uttermost parts of the earth in the whirlwind of his wrath

M**

Mind, I am not saying his whisky ever affected his
head or his legs, for it did not, in even the slightest
degree. He was a capacious container, but he did
not hold enough for that. He took a level tumbler-
ful of whisky every morning before he put his
clothes on —"to sweeten his bilgewater," he said.
He took another after he got the most of his
clothes on, "to settle his mind and give him his
bearings." He then shaved, and put on a clean
shirt; after which he recited the Lord's Prayer in a
fervent, thundering bass that shook the ship to her
kelson and suspended all conversation in the main
cabin. Then, at this stage, being invariably "by
the head," or "by the stern," or "listed to port or
starboard," he took one more to "put him on an
even keel so that he would mind his hellum and not
miss stays and go about, every time he came up in
the wind." And now, his state-room door swung
open and the sun of his benignant face beamed redly
out upon men and women and children, and he
roared his "Shipmets a'hoy!" in a way that was
calculated to wake the dead and precipitate the final
resurrection; and forth he strode, a picture to look
at and a presence to enforce attention. Stalwart
and portly; not a gray hair; broad-brimmed slouch
hat; semi-sailor toggery of blue navy flannel —
roomy and ample; a stately expanse of shirt-front
and a liberal amount of black silk neckcloth tied
with a sailor knot; large chain and imposing seals
impending from his fob; awe-inspiring feet, and "a

hand like the hand of Providence," as his whaling brethren expressed it; wristbands and sleeves pushed back half way to the elbow, out of respect for the warm weather, and exposing hairy arms, gaudy with red and blue anchors, ships, and goddesses of liberty tattooed in India ink. But these details were only secondary matters — his face was the lodestone that chained the eye. It was a sultry disk, glowing determinedly out through a weather-beaten mask of mahogany, and studded with warts, seamed with scars, "blazed" all over with unfailing fresh slips of the razor; and with cheery eyes, under shaggy brows, contemplating the world from over the back of a gnarled crag of a nose that loomed vast and lonely out of the undulating immensity that spread away from its foundations. At his heels frisked the darling of his bachelor estate, his terrier "Fan," a creature no larger than a squirrel. The main part of his daily life was occupied in looking after "Fan," in a motherly way, and doctoring her for a hundred ailments which existed only in his imagination.

The Admiral seldom read newspapers; and when he did he never believed anything they said. He read nothing, and believed in nothing, but "The Old Guard," a secession periodical published in New York. He carried a dozen copies of it with him, always, and referred to them for all required information. If it was not there, he supplied it himself, out of a bountiful fancy, inventing history,

names, dates, and everything else necessary to make
his point good in an argument. Consequently, he
was a formidable antagonist in a dispute. Whenever he swung clear of the record and began to
create history, the enemy was helpless and had to
surrender. Indeed, the enemy could not keep
from betraying some little spark of indignation at
his manufactured history — and when it came to indignation, that was the Admiral's very "best hold."
He was always ready for a political argument, and
if nobody started one he would do it himself. With
his third retort his temper would begin to rise, and
within five minutes he would be blowing a gale, and
within fifteen his smoking-room audience would be
utterly stormed away and the old man left solitary
and alone, banging the table with his fist, kicking
the chairs, and roaring a hurricane of profanity. It
got so, after a while, that whenever the Admiral approached, with politics in his eye, the passengers
would drop out with quiet accord, afraid to meet
him; and he would camp on a deserted field.

But he found his match at last, and before a full
company. At one time or another, everybody had
entered the lists against him and been routed, except the quiet passenger Williams. He had never
been able to get an expression of opinion out of him
on politics. But now, just as the Admiral drew near
the door and the company were about to slip out,
Williams said:

"Admiral, are you *certain* about that circumstance concerning the clergymen you mentioned the other day?''— referring to a piece of the Admiral's manufactured history.

Every one was amazed at the man's rashness. The idea of deliberately inviting annihilation was a thing incomprehensible. The retreat came to a halt; then everybody sat down again wondering, to await the upshot of it. The Admiral himself was as surprised as any one. He paused in the door, with his red handkerchief half raised to his sweating face, and contemplated the daring reptile in the corner.

"*Certain* of it? Am I *certain* of it? Do you think I've been lying about it? What do you take me for? Anybody that don't know that circumstance, don't know anything; a child ought to know it. Read up your history! Read it up —— —— —— ——, and don't come asking a man if he's *certain* about a bit of A B C stuff that the very Southern niggers know all about."

Here the Admiral's fires began to wax hot, the atmosphere thickened, the coming earthquake rumbled, he began to thunder and lighten. Within three minutes his volcano was in full irruption and he was discharging flames and ashes of indignation, belching black volumes of foul history aloft, and vomiting red-hot torrents of profanity from his crater. Meantime Williams sat silent, and apparently deeply and earnestly interested in what the old man was saying. By and by, when the lull came, he said in the most deferential way, and with the

gratified air of a man who has had a mystery cleared
up which had been puzzling him uncomfortably:

"*Now*, I understand it. I always thought I
knew that piece of history well enough, but was still
afraid to trust it, because there was not that convincing particularity about it that one likes to have
in history; but when you mentioned every name,
the other day, and every date, and every little circumstance, in their just order and sequence, I said
to myself, *this* sounds something like — *this* is
history — *this* is putting it in a shape that gives a
man confidence; and I said to myself afterward, I
will just ask the **Admiral** if he is perfectly certain
about the details, and if he is I will come out and
thank him for clearing this matter up for me. And
that is what I want to do now — for until you set
that matter right it was nothing but just a confusion
in my mind, without head or tail to it."

Nobody ever saw the **Admiral** look so mollified
before, and so pleased. Nobody had ever received
his bogus history as gospel before; its genuineness
had always been called in question either by words
or looks; but here was a man that not only swallowed it all down, but was grateful for the dose.
He was taken aback; he hardly knew what to say;
even his profanity failed him. Now, Williams continued, modestly and earnestly:

"But, Admiral, in saying that this was the first
stone thrown, and that this precipitated the war,
you have overlooked a circumstance which you are

perfectly familiar with, but which has escaped your memory. Now I grant you that what you have stated is correct in every detail — to wit: that on the 16th of October, 1860, two Massachusetts clergymen, named Waite and Granger, went in disguise to the house of John Moody, in Rockport, at dead of night, and dragged forth two Southern women and their two little children, and after tarring and feathering them conveyed them to Boston and burned them alive in the State House square; and I also grant your proposition that this deed is what led to the secession of South Carolina on the 20th of December following. Very well." [Here the company were pleasantly surprised to hear Williams proceed to come back at the Admiral with his own invincible weapon — clean, pure, *manufactured history*, without a word of truth in it.] "Very well, I say. But, Admiral, why overlook the Willis and Morgan case in South Carolina? You are too well informed a man not to know all about that circumstance. Your arguments and your conversations have shown you to be intimately conversant with every detail of this national quarrel. You develop matters of history every day that show plainly that you are no smatterer in it, content to nibble about the surface, but a man who has searched the depths and possessed yourself of everything that has a bearing upon the great question. Therefore, let me just recall to your mind that Willis and Morgan case — though I see by your face that the whole thing is

already passing through your memory at this mo‧
ment. On the 12th of August, 1860, *two months*
before the Waite and Granger affair, two South
Carolina clergymen, named John H. Morgan and
Winthrop L. Willis, one a Methodist and the other
an Old School Baptist, disguised themselves, and
went at midnight to the house of a planter named
Thompson — Archibald F. Thompson, vice-president under Thomas Jefferson,— and took thence, at
midnight, his widowed aunt (a Northern woman),
and her adopted child, an orphan named Mortimer
Highie, afflicted with epilepsy and suffering at the
time from white swelling on one of his legs, and
compelled to walk on crutches in consequence; and
the two ministers, in spite of the pleadings of the
victims, dragged them to the bush, tarred and
feathered them, and afterward burned them at the
stake in the city of Charleston. You remember
perfectly well what a stir it made; you remember
perfectly well that even the Charleston *Courier* stig-
matized the act as being unpleasant, of questionable
propriety, and scarcely justifiable, and likewise that
it would not be matter of surprise if retaliation ensued. And you remember also, that this thing was
the *cause* of the Massachusetts outrage. Who,
indeed, were the two Massachusetts ministers? and
who were the two Southern women they burned? I
do not need to remind *you*, admiral, with your intimate knowledge of history, that Waite was the
nephew of the woman burned in Charleston; that

Granger was her cousin in the second degree, and
that the woman they burned in Boston was the wife
of John H. Morgan, and the still loved but divorced
wife of Winthrop L. Willis. Now, Admiral, it is
only fair that you should acknowledge that the first
provocation came from the Southern preachers and
that the Northern ones were justified in retaliating.
In your arguments you never yet have shown the
least disposition to withhold a just verdict or be in
anywise unfair, when authoritative history condemned your position, and therefore I have no hesi-
tation in asking you to take the original blame from
the Massachusetts ministers, in this matter, and
transfer it to the South Carolina clergymen where it
justly belongs."

The Admiral was conquered. This sweet-spoken
creature who swallowed his fraudulent history as if
it were the bread of life; basked in his furious
blasphemy as if it were generous sunshine; found
only calm, even-handed justice in his rampart
partisanship; and flooded him with invented history
so sugar-coated with flattery and deference that
there was no rejecting it, was "too many" for him.
He stammered some awkward, profane sentences
about the ―― ―― ―― ―― Willis and Morgan
business having escaped his memory, but that he
"remembered it now," and then, under pretense
of giving Fan some medicine for an imaginary
cough, drew out of the battle and went away, a
vanquished man. Then cheers and laughter went

up, and williams, the ship's benefactor, was a nero.
The news went about the vessel, champagne was
ordered, an enthusiastic reception instituted in the
smoking-room, and everybody flocked thither to
shake hands with the conqueror. The wheelsman
said afterward, that the Admiral stood up behind the
pilot house and "ripped and cursed all to himself"
till he loosened the smoke-stack guys and becalmed
the mainsail.

The Admiral's power was broken. After that, if
he began an argument, somebody would bring
Williams, and the old man would grow weak and
begin to quiet down at once. And as soon as he
was done, Williams in his dulcet, insinuating way,
would invent some history (referring for proof, to
the old man's own excellent memory and to copies
of "The Old Guard" known not to be in his possession) that would turn the tables completely and
leave the Admiral all abroad and helpless. By and
by he came to so dread Williams and his gilded
tongue that he would stop talking when he saw him
approach, and finally ceased to mention politics
altogether, and from that time forward there was
entire peace and serenity in the ship.

CHAPTER XXII.

ON a certain bright morning the Islands hove in sight, lying low on the lonely sea, and everybody climbed to the upper deck to look. After two thousand miles of watery solitude the vision was a welcome one. As we approached, the imposing promontory of Diamond Head rose up out of the ocean, its rugged front softened by the hazy distance, and presently the details of the land began to make themselves manifest: first the line of beach; then the plumed cocoanut trees of the tropics; then cabins of the natives; then the white town of Honolulu, said to contain between twelve and fifteen thousand inhabitants, spread over a dead level; with streets from twenty to thirty feet wide, solid and level as a floor, most of them straight as a line and few as crooked as a corkscrew.

The further I traveled through the town the better I liked it. Every step revealed a new contrast — disclosed something I was unaccustomed to. In place of the grand mud-colored brown fronts of San Francisco, I saw dwellings built of straw, adobes, and cream-colored pebble-and-shell-conglomerated coral,

cut into oblong blocks and laid in cement; also a
great number of neat white cottages, with green
window-shutters; in place of front yards like bil-
liard-tables with iron fences around them, I saw these
homes surrounded by ample yards, thickly clad with
green grass, and shaded by tall trees, through whose
dense foliage the sun could scarcely penetrate; in
place of the customary geranium, calla lily, etc.,
languishing in dust and general debility, I saw luxurious banks and thickets of flowers, fresh as a
meadow after a rain, and glowing with the richest
dyes; in place of the dingy horrors of San Francisco's pleasure grove, the "Willows," I saw huge-
bodied, wide-spreading forest trees, with strange
names and stranger appearance — trees that cast a
shadow like a thundercloud, and were able to stand
alone without being tied to green poles; in place of
gold fish, wiggling around in glass globes, assuming
countless shades and degrees of distortion through
the magnifying and diminishing qualities of their
transparent prison-house, I saw cats — Tom cats,
Mary Ann cats, long-tailed cats, bob-tailed cats,
blind cats, one-eyed cats, wall-eyed cats, cross-eyed
cats, gray cats, black cats, white cats, yellow cats,
striped cats, spotted cats, tame cats, wild cats,
singed cats, individual cats, groups of cats, platoons
of cats, companies of cats, regiments of cats,
armies of cats, multitudes of cats, millions of cats,
and all of them sleek, fat, lazy, and sound asleep.

I looked on a multitude of people, some white,

in white coats, vests, pantaloons, even white cloth
shoes, made snowy with chalk duly laid on every
morning; but the majority of the people were almost
as dark as negroes — women with comely features,
fine black eyes, rounded forms, inclining to the
voluptuous, clad in a single bright red or white
garment that fell free and unconfined from shoulder
to heel, long black hair falling loose, gypsy hats,
encircled with wreaths of natural flowers of a brilliant carmine tint; plenty of dark men in various
costumes, and some with nothing on but a battered
stove-pipe hat tilted on the nose, and a very scant
breech-clout; certain smoke - dried children were
clothed in nothing but sunshine — a very neat fitting
and picturesque apparel indeed.

In place of roughs and rowdies staring and black-
guarding on the corners, I saw long-haired, saddle-
colored Sandwich Island maidens sitting on the
ground in the shade of corner houses, gazing indo-
lently at whatever or whoever happened along; instead of wretched cobble-stone pavements, I walked
on a firm foundation of coral, built up from the bottom of the sea by the absurd but persevering insect
of that name, with a light layer of lava and cinders
overlying the coral, belched up out of fathomless
perdition long ago through the seared and blackened
crater that stands dead and harmless in the distance
now; instead of cramped and crowded street-cars,
I met dusky native women sweeping by, free as the
wind, on fleet horses and astride, with gaudy riding-

sashes, streaming like banners behind them; instead
of the combined stenches of Chinadom and Brannan
street slaughter-houses, I breathed the balmy
fragrance of jessamine, oleander, and the Pride of
India; in place of the hurry and bustle and noisy
confusion of San Francisco, I moved in the midst
of a summer calm as tranquil as dawn in the Garden
of Eden; in place of the Golden City's skirting sand-
hills and the placid bay, I saw on the one side a
frame-work of tall, precipitous mountains close at
hand, clad in refreshing green, and cleft by deep,
cool, chasm-like valleys — and in front the grand
sweep of the ocean: a brilliant, transparent green
near the shore, bound and bordered by a long white
line of foamy spray dashing against the reef, and
further out the dead blue water of the deep sea,
flecked with "white caps," and in the far horizon
a single, lonely sail — a mere accent-mark to em-
phasize a slumberous calm and a solitude that were
without sound or limit. When the sun sunk down
— the one intruder from other realms and persistent
in suggestions of them — it was tranced luxury to
sit in the perfumed air and forget that there was any
world but these enchanted islands.

It was such ecstasy to dream and dream — till you
got a bite. A scorpion bite. Then the first duty
was to get up out of the grass and kill the scorpion;
and the next to bathe the bitten place with alcohol
or brandy; and the next to resolve to keep out of
the grass in the future. Then came an adjourn-

ment to the bed-chamber and the pastime of writing up the day's journal with one hand and the destruction of mosquitoes with the other — a whole community of them at a slap. Then, observing an enemy approaching,— a hairy tarantula on stilts — why not set the spittoon on him? It is done, and the projecting ends of his paws give a luminous idea of the magnitude of his reach. Then to bed and become a promenade for a centipede with forty-two legs on a side and every foot hot enough to burn a hole through a rawhide. More soaking with alcohol, and a resolution to examine the bed before entering it, in future. Then wait, and suffer, till all the mosquitoes in the neighborhood have crawled in under the bar, then slip out quickly, and shut them in and sleep peacefully on the floor till morning. Meantime it is comforting to curse the tropics in occasional wakeful intervals.

We had an abundance of fruit in Honolulu, of course. Oranges, pine-apples, bananas, strawberries, lemons, limes, mangoes, guavas, melons, and a rare and curious luxury called the chirimoya, which is deliciousness itself. Then there is the tamarind. I thought tamarinds were made to eat, but that was probably not the idea. I ate several, and it seemed to me that they were rather sour that year. They pursed up my lips, till they resembled the stem-end of a tomato, and I had to take my sustenance through a quill for twenty-four hours. They sharpened my teeth till I could have shaved with

14.

them, and gave them a "wire edge" that I was
afraid would stay; but a citizen said "no, it will
come off when the enamel does" — which was comforting, at any rate. I found, afterward, that only
strangers eat tamarinds — but they only eat them
once.

CHAPTER XXIII.

IN my diary of our third day in Honolulu, I find this:

I am probably the most sensitive man in Hawaii to-night — especially about sitting down in the presence of my betters. I have ridden fifteen or twenty miles on horseback since 5 P.M., and to tell the honest truth, I have a delicacy about sitting down at all.

An excursion to Diamond Head and the King's Cocoanut Grove was planned to-day — time 4.30 P.M. — the party to consist of half a dozen gentlemen and three ladies. They all started at the appointed hour except myself. I was at the government prison (with Captain Fish and another whaleship-skipper, Captain Phillips), and got so interested in its examination that I did not notice how quickly the time was passing. Somebody remarked that it was twenty minutes past five o'clock, and that woke me up. It was a fortunate circumstance that Captain Phillips was along with his "turn out," as he calls a top-buggy that Captain Cook brought here in 1778, and a horse that was here when Captain Cook came. Captain Phillips takes a just pride in his

N**

driving and in the speed of his horse, and to his passion for displaying them I owe it that we were only
sixteen minutes coming from the prison to the
American Hotel — a distance which has been estimated to be over half a mile. But it took some
fearful driving. The Captain's whip came down
fast, and the blows started so much dust out of the
horse's hide that during the last half of the journey
we rode through an impenetrable fog, and ran by
a pocket compass in the hands of Captain Fish, a
whaler of twenty-six years' experience, who sat
there through the perilous voyage as self-possessed
as if he had been on the euchre-deck of his own ship,
and calmly said, " Port your helm — port," from
time to time, and " Hold her a little free — steady
— so-o," and " Luff — hard down to starboard ! "
and never once lost his presence of mind or betrayed
the least anxiety by voice or manner. When we
came to anchor at last, and Captain Phillips looked
at his watch and said, " Sixteen minutes — I told
you it was in her ! that's over three miles an hour ! "
I could see he felt entitled to a compliment, and
so I said I had never seen lightning go like that
horse. And I never had.

The landlord of the American said the party had
been gone nearly an hour, but that he could give
me my choice of several horses that could overtake
them. I said, never mind — I preferred a safe
horse to a fast one — I would like to have an excessively gentle horse — a horse with no spirit whatever

— a lame one, if he had such a thing. Inside of five minutes I was mounted, and perfectly satisfied with my outfit. I had no time to label him "This is a horse," and so if the public took him for a sheep I cannot help it. I was satisfied, and that was the main thing. I could see that he had as many fine points as any man's horse, and so I hung my hat on one of them, behind the saddle, and swabbed the perspiration from my face and started. I named him after this island, "Oahu" (pronounced O-waw-hee). The first gate he came to he started in; I had neither whip nor spur, and so I simply argued the case with him. He resisted argument, but ultimately yielded to insult and abuse. He backed out of that gate and steered for another one on the other side of the street. I triumphed by my former process. Within the next six hundred yards he crossed the street fourteen times and attempted thirteen gates, and in the meantime the tropical sun was beating down and threatening to cave the top of my head in, and I was literally dripping with perspiration. He abandoned the gate business after that and went along peaceably enough, but absorbed in meditation. I noticed this latter circumstance, and it soon began to fill me with apprehension. I said to myself, this creature is planning some new outrage, some fresh deviltry or other — no horse ever thought over a subject so profoundly as this one is doing just for nothing. The more this thing preyed upon my mind the more uneasy I became, until the sus-

pense became almost unbearable, and I dismounted
to see if there was anything wild in his eye — for I
had heard that the eye of this noblest of our domes-
tic animals is very expressive. I cannot describe
what a load of anxiety was lifted from my mind when
I found that he was only asleep. I woke him up
and started him into a faster walk, and then the villainy of his nature came out again. He tried to
climb over a stone wall, five or six feet high. I
saw that I must apply force to this horse, and that
I might as well begin first as last. I plucked a stout
switch from a tamarind tree, and the moment he saw
it, he surrendered. He broke into a convulsive sort
of a canter, which had three short steps in it and one
long one, and reminded me alternately of the clattering
shake of the great earthquake, and the sweeping
plunge of the *Ajax* in a storm.

And now there can be no fitter occasion than the
present to pronounce a left-handed blessing upon
the man who invented the American saddle. There
is no seat to speak of about it — one might as well
sit in a shovel — and the stirrups are nothing but an
ornamental nuisance. If I were to write down here
all the abuse I expended on those stirrups, it would
make a large book, even without pictures. Sometimes I got one foot so far through, that the stirrup
partook of the nature of an anklet; sometimes both
feet were through, and I was handcuffed by the
legs; and sometimes my feet got clear out and
left the stirrups wildly dangling about my shins,

Even when I was in proper position and carefully
balanced upon the balls of my feet, there was no
comfort in it, on account of my nervous dread that
they were going to slip one way or the other in a
moment.   But the subject is too exasperating to
write about.

A mile and a half from town, I came to a grove of
tall cocoanut trees, with clean, branchless stems
reaching straight up sixty or seventy feet and topped
with a spray of green foliage sheltering clusters of
cocoanuts — not more picturesque than a forest of
colossal ragged parasols, with bunches of magnified
grapes under them, would be.   I once heard a
grouty Northern invalid say that a cocoanut tree
might be poetical, possibly it was; but it looked
like a feather-duster struck by lightning.   I think
that describes it better than a picture — and yet,
without any question, there is something fascinating
about a cocoanut tree — and graceful, too.

About a dozen cottages, some frame and the others
of native grass, nestled sleepily in the shade here
and there.   The grass cabins are of a grayish color,
are shaped much like our own cottages, only with
higher and steeper roofs, usually, and are made of
some kind of weed strongly bound together in
bundles.   The roofs are very thick, and so are the
walls, the latter have square holes in them for windows.   At a little distance these cabins have a furry
appearance, as if they might be made of bear
skins.   They are very cool and pleasant inside.   The

King's flag was flying from the roof of one of the cottages, and His Majesty was probably within. He owns the whole concern thereabouts, and passes his time there frequently, on sultry days "laying off." The spot is called "The King's Grove."

Near by is an interesting ruin — the meager remains of an ancient temple — a place where human sacrifices were offered up in those old bygone days when the simple child of nature, yielding momentarily to sin when sorely tempted, acknowleged his error when calm reflection had shown it to him, and came forward with noble frankness and offered up his grandmother as an atoning sacrifice — in those old days when the luckless sinner could keep on cleansing his conscience and achieving periodical happiness as long as his relations held out; long, long before the missionaries braved a thousand privations to come and make them permanently miserable by telling them how beautiful and how blissful a place heaven is, and how nearly impossible it is to get there; and showed the poor native how dreary a place perdition is and what unnecessarily liberal facilities there are for going to it; showed him how, in his ignorance, he had gone and fooled away all his kinsfolk to no purpose; showed him what rapture it is to work all day long for fifty cents to buy food for next day with, as compared with fishing for a pastime and lolling in the shade through eternal summer, and eating of the bounty that nobody labored to provide but Nature. How sad it is to think of the multitudes who have gone to their graves in this beautiful island and never knew there was a hell.

This ancient temple was built of rough blocks of lava, and was simply a roofless inclosure a hundred and thirty feet long and seventy wide — nothing but naked walls, very thick, but not much higher than a man's head. They will last for ages, no doubt, if left unmolested. Its three altars and other sacred appurtenances have crumbled and passed away years ago. It is said that in the old times thousands of human beings were slaughtered here, in the presence of naked and howling savages. If these mute stones could speak, what tales they could tell, what pictures they could describe, of fettered victims writhing under the knife; of massed forms straining forward out of the gloom, with ferocious faces lit up by the sacrificial fires; of the background of ghostly trees; of the dark pyramid of Diamond Head standing sentinel over the uncanny scene, and the peaceful moon looking down upon it through rifts in the cloud-rack!

When Kamehameha (pronounced Ka-may-ha-may-ah) the Great — who was a sort of a Napoleon in military genius and uniform success — invaded this island of Oahu three-quarters of a century ago, and exterminated the army sent to oppose him, and took full and final possession of the country, he searched out the dead body of the King of Oahu, and those of the principal chiefs, and impaled their heads on the walls of this temple.

Those were savage times when this old slaughter-house was in its prime. The King and the chiefs ruled the common herd with a rod of iron; made them gather all the provisions the masters needed; build all the houses and temples; stand all the expenses, of whatever kind; take kicks and cuffs for thanks; drag out lives well flavored with misery, and then suffer death for trifling offenses or yield up their lives on the sacrificial altars to purchase favors from the gods for their hard rulers. The missionaries have clothed them, educated them, broken up the tyrannous authority of their chiefs, and given them freedom and the right to enjoy whatever their hands and brains produce, with equal laws for all, and punishment for all alike who transgress them. The contrast is so strong — the benefit conferred upon this people by the missionaries is so prominent, so palpable, and so unquestionable, that the frankest compliment I can pay them, and the best, is simply to point to the condition of the Sandwich Islanders of Captain Cook's time, and their condition to-day. Their work speaks for itself.

CHAPTER XXIV.

BY and by, after a rugged climb, we halted on the summit of a hill which commanded a far-reaching view. The moon rose and flooded mountain and valley and ocean with a mellow radiance, and out of the shadows of the foliage the distant lights of Honolulu glinted like an encampment of fireflies. The air was heavy with the fragrance of flowers. The halt was brief. Gayly laughing and talking, the party galloped on, and I clung to the pommel and cantered after. Presently we came to a place where no grass grew — a wide expanse of deep sand. They said it was an old battle ground. All around everywhere, not three feet apart, the bleached bones of men gleamed white in the moonlight. We picked up a lot of them for mementoes. I got quite a number of arm bones and leg bones — of great chiefs, may be, who had fought savagely in that fearful battle in the old days, when blood flowed like wine where we now stood,—and wore the choicest of them out on Oahu afterward, trying to make him go. All sorts of bones could be found except skulls; but a citizen said, irreverently, that there had been an unusual number of "skull-hunters" there lately — a species of sportsmen I had never heard of before.

Nothing whatever is known about this place — its story is a secret that will never be revealed. The oldest natives make no pretense of being possessed of its history. They say these bones were here when they were children. They were here when their grandfathers were children — but how they came here, they can only conjecture. Many people believe this spot to be an ancient battle-ground, and it is usual to call it so; and they believe that these skeletons have lain for ages just where their proprietors fell in the great fight. Other people believe that Kamehameha I. fought his first battle here. On this point, I have heard a story, which may have been taken from one of the numerous books which have been written concerning these islands — I do not know where the narrator got it. He said that when Kamehameha (who was at first merely a subordinate chief on the island of Hawaii), landed here, he brought a large army with him, and encamped at Waikiki. The Oahuans marched against him, and so confident were they of success that they readily acceded to a demand of their priests that they should draw a line where these bones now lie, and take an oath that, if forced to retreat at all, they would never retreat beyond this boundary. The priests told them that death and everlasting punishment would overtake any who violated the oath, and

the march was resumed. Kamehameha drove them
back step by step; the priests fought in the front
rank and exhorted them both by voice and inspiriting example to remember their oath — to die, if need
be, but never cross the fatal line. The struggle was
manfully maintained, but at last the chief priest fell,
pierced to the heart with a spear, and the unlucky
omen fell like a blight upon the brave souls at his
back; with a triumphant shout the invaders pressed
forward — the line was crossed — the offended gods
deserted the despairing army, and, accepting the
doom their perjury had brought upon them, they
broke and fled over the plain where Honolulu stands
now — up the beautiful Nuuanu Valley — paused a
moment, hemmed in by precipitous mountains on
either hand and the frightful precipice of the Pari in
front, and then were driven over — a sheer plunge
of six hundred feet!

The story is pretty enough, but Mr. Jarves' excellent history says the Oahuans were intrenched in
Nuuanu Valley; that Kamehameha ousted them,
routed them, pursued them up the valley and drove
them over the precipice. He makes no mention of
our bone-yard at all in his book.

Impressed by the profound silence and repose that
rested over the beautiful landscape, and being, as
usual, in the rear, I gave voice to my thoughts. I
said:

"What a picture is here slumbering in the solemn
glory of the moon! How strong the rugged outlines

of the dead volcano stand out against the clear sky!
What a snowy fringe marks the bursting of the surf
over the long, curved reef! How calmly the dim
city sleeps yonder in the plain! How soft the
shadows lie upon the stately mountains that border
the dream-haunted Mauoa Valley! What a grand
pyramid of billowy clouds towers above the storied
Pari! How the grim warriors of the past seem flock-
ing in ghostly squadrons to their ancient battlefield
again — how the wails of the dying well up from
the ——''

At this point the horse called Oahu sat down in the
sand. Sat down to listen, I suppose. Never mind
what he heard, I stopped apostrophizing and convinced him that I was not a man to allow contempt
of court on the part of a horse. I broke the backbone of a chief over his rump and set out to join the
cavalcade again.

Very considerably fagged out we arrived in town
at 9 o'clock at night, myself in the lead — for when
my horse finally came to understand that he was
homeward bound and hadn't far to go, he turned
his attention strictly to business.

This is a good time to drop in a paragraph of information. There is no regular livery - stable in
Honolulu, or, indeed, in any part of the kingdom of
Hawaii; therefore unless you are acquainted with
wealthy residents (who all have good horses), you
must hire animals of the wretchedest description
from the Kanakas (*i. e.*, natives). Any horse you

hire, even though it be from a white man, is not
often of much account, because it will be brought
in for you from some ranch, and has necessarily been
leading a hard life. If the Kanakas who have been
caring for him (inveterate riders they are) have not
ridden him half to death every day themselves, you
can depend upon it they have been doing the same
thing by proxy, by clandestinely hiring him out.
At least, so I am informed. The result is, that no
horse has a chance to eat, drink, rest, recuperate, or
look well or feel well, and so strangers go about the
Islands mounted as I was to-day.

In hiring a horse from a Kanaka, you must have
all your eyes about you, because you can rest satisfied that you are dealing with a shrewd, unprincipled
rascal. You may leave your door open and your
trunk unlocked as long as you please, and he will
not meddle with your property; he has no important vices and no inclination to commit robbery
on a large scale; but if he can get ahead of you in
the horse business, he will take a genuine delight in
doing it. This trait is characteristic of horse jockeys,
the world over, is it not? He will overcharge you
if he can: he will hire you a fine-looking horse at
night (anybody's — may be the King's, if the royal
steed be in convenient view), and bring you the mate
to my Oahu in the morning, and contend that it is
the same animal. If you make trouble, he will get out
by saying it was not himself who made the bargain
with you, but his brother, "who went out in the

country this morning." They have always got a "brother" to shift the responsibility upon. A victim said to one of these fellows one day:

"But I know I hired the horse of you, because I noticed that scar on your cheek."

The reply was not bad: "Oh, yes — yes — my brother all same — we twins!"

A friend of mine, J. Smith, hired a horse yesterday, the Kanaka warranting him to be in excellent condition. Smith had a saddle and blanket of his own, and he ordered the Kanaka to put these on the horse. The Kanaka protested that he was perfectly willing to trust the gentleman with the saddle that was already on the animal, but Smith refused to use it. The change was made: then Smith noticed that the Kanaka had only changed the saddles, and had left the original blanket on the horse; he said he forgot to change the blankets, and so, to cut the bother short, Smith mounted and rode away. The horse went lame a mile from town, and afterward got to cutting up some extraordinary capers. Smith got down and took off the saddle, but the blanket stuck fast to the horse — glued to a procession of raw places. The Kanaka's mysterious conduct stood explained.

Another friend of mine bought a pretty good horse from a native, a day or two ago, after a tolerably thorough examination of the animal. He discovered to-day that the horse was as blind as a bat, in one eye. He meant to have examined that eye, and

came home with a general notion that he had done
it; but he remembered now that every time he made
the attempt his attention was called to something else
by his victimizer.

One more instance, and then I will pass to something else. I am informed that when a certain Mr.
L., a visiting stranger, was here, he bought a pair
of very respectable-looking match horses from a
native. They were in a little stable with a partition
through the middle of it — one horse in each apartment. Mr. L. examined one of them critically
through a window (the Kanaka's "brother" having
gone to the country with the key), and then went
around the house and examined the other through
a window on the other side. He said it was the
neatest match he had ever seen, and paid for the
horses on the spot. Whereupon the Kanaka departed to join his brother in the country. The fellow had shamefully swindled L. There was only
one "match" horse, and he had examined his starboard side through one window and his port side
through another! I decline to believe this story, but
I give it because it is worth something as a fanciful
illustration of a fixed fact — namely, that the Kanaka horse-jockey is fertile in invention and elastic in
conscience.

You can buy a pretty good horse for forty or fifty
dollars, and a good enough horse for all practical
purposes for two dollars and a half. I estimate
"Oahu" to be worth somewhere in the neighborhood

15**

of thirty-five cents. A good deal better animal than
he is was sold here day before yesterday for a dollar
and seventy-five cents, and sold again to-day for two
dollars and twenty-five cents; Williams bought a
handsome and lively little pony yesterday for ten
dollars; and about the best common horse on the
island (and he is a really good one) sold yesterday,
with Mexican saddle and bridle, for seventy dollars
— a horse which is well and widely known, and
greatly respected for his speed, good disposition, and
everlasting bottom. You give your horse a little
grain once a day; it comes from San Francisco, and
is worth about two cents a pound; and you give
him as much hay as he wants; it is cut and brought
to the market by natives, and is not very good; it
is baled into long, round bundles, about the size of a
large man; one of them is stuck by the middle on
each end of a six-foot pole, and the Kanaka shoulders
the pole and walks about the streets between the upright bales in search of customers. These hay bales,
thus carried, have a general resemblance to a colossal
capital H.

The hay-bundles cost twenty-five cents apiece, and
one will last a horse about a day. You can get a
horse for a song, a week's hay for another song, and
you can turn your animal loose among the luxuriant
grass in your neighbor's broad front yard without a
song at all — you do it at midnight, and stable the
beast again before morning. You have been at no
expense thus far, but when you come to buy a sad-

dle and bridle they will cost you from twenty to thirty-five dollars. You can hire a horse, saddle, and bridle at from seven to ten dollars a week, and the owner will take care of them at his own expense.

It is time to close this day's record — bedtime. As I prepare for sleep, a rich voice rises out of the still night, and, far as this ocean rock is toward the ends of the earth, I recognize a familiar home air. But the words seem somewhat out of joint:

"Waikiki lantoni œ Kaa hocly hooly wawhoo."

Translated, that means "When we were marching through Georgia."

CHAPTER XXV.

PASSING through the market-place we saw that feature of Honolulu under its most favorable auspices — that is, in the full glory of Saturday afternoon, which is a festive day with the natives. The native girls, by twos and threes and parties of a dozen, and sometimes in whole platoons and companies, went cantering up and down the neighboring streets astride of fleet but homely horses, and with their gaudy riding-habits streaming like banners behind them. Such a troop of free and easy riders, in their natural home, the saddle, makes a gay and graceful spectacle. The riding-habit I speak of is simply a long, broad scarf, like a tavern tablecloth, brilliantly colored, wrapped around the loins once, then apparently passed between the limbs and each end thrown backward over the same, and floating and flapping behind on both sides beyond the horse's tail like a couple of fancy flags; then, slipping the stirrup-irons between her toes, the girl throws her chest forward, sits up like a major-general, and goes sweeping by like the wind.

The girls put on all the finery they can on Satur-

day afternoon — fine black silk robes; flowing red
ones that nearly put your eyes out; others as white
as snow; still others that discount the rainbow; and
they wear their hair in nets, and trim their jaunty hats
with fresh flowers, and encircle their dusky throats
with home-made necklaces of the brilliant vermilion-
tinted blossom of the *ohia;* and they fill the markets
and the adjacent streets with their bright presences,
and smell like a rag factory on fire with their offensive
cocoanut oil.

Occasionally, you see a heathen from the sunny
isles away down in the South Seas, with his face
and neck tattooed till he looks like the customary
mendicant from Washoe who has been blown up in
a mine.   Some are tattooed a dead blue color down
to the upper lip — masked, as it were — leaving the
natural light yellow skin of Micronesia unstained
from thence down; some with broad marks drawn
down from hair to neck, on both sides of the face,
and a strip of the original yellow skin, two inches
wide, down the center — a gridiron with a spoke
broken out; and some with the entire face discolored
with the popular mortification tint, relieved only by
one or two thin, wavy threads of natural yellow running across the face from ear to ear, and eyes twinkling out of this darkness, from under shadowing hat-
brims, like stars in the dark of the moon.

Moving among the stirring crowds, you come to the
poi merchants, squatting in the shade on their hams,
in true native fashion, and surrounded by purchasers.

(The Sandwich Islanders always squat on their hams,
and who knows but they may be the original " ham
sandwiches"? The thought is pregnant with interest.) The poi looks like common flour paste, and
is kept in large bowls formed of a species of gourd,
and capable of holding from one to three or four
gallons. Poi is the chief article of food among the
natives, and is prepared from the *taro* plant. The
taro root looks like a thick, or, if you please, a cor-
pulent sweet potato, in shape, but is of a light pur-
ple color when boiled. When boiled it answers as a
passable substitute for bread. The buck Kanakas
bake it under ground, then mash it up well with a
heavy lava pestle, mix water with it until it becomes
a paste, set it aside and let it ferment, and then it is
poi — and an unseductive mixture it is, almost taste-
less before it ferments and too sour for a luxury
afterward. But nothing is more nutritious. When
solely used, however, it produces acrid humors, a
fact which sufficiently accounts for the humorous
character of the Kanakas. I think there must be as
much of a knack in handling poi as there is in eating
with chopsticks. The forefinger is thrust into the
mess and stirred quickly round several times and
drawn as quickly out, thickly coated, just as if it
were poulticed; the head is thrown back, the finger
inserted in the mouth and the delicacy stripped off
and swallowed — the eye closing gently, meanwhile,
in a languid sort of ecstasy. Many a different
finger goes into the same bowl and many a different

kind of dirt and shade and quality of flavor is added
to the virtues of its contents.

Around a small shanty was collected a crowd of
natives buying the *awa* root.   It is said that but for
the use of this root the destruction of the people in
former times by certain imported diseases would
have been far greater than it was, and by others it is
said that this is merely a fancy.   All agree that poi
will rejuvenate a man who is used up and his vitality
almost annihilated by hard drinking, and that in
some kinds of diseases it will restore health after all
medicines have failed; but all are not willing to allow
to the *awa* the virtues claimed for it.   The natives
manufacture an intoxicating drink from it which is
fearful in its effects when persistently indulged in.
It covers the body with dry, white scales, inflames
the eyes, and causes premature decrepitude.
Although the man before whose establishment we
stopped has to pay a government license of eight
hundred dollars a year for the exclusive right to sell
awa root, it is said that he makes a small fortune
every twelve-month; while saloon-keepers, who pay
a thousand dollars a year for the privilege of retail-
ing whisky, etc., only make a bare living.

We found the fish market crowded; for the native
is very fond of fish, and *eats the article raw and
alive!*   Let us change the subject.

In old times here Saturday was a grand gala day
indeed.   All the native population of the town forsook their labors, and those of the surrounding

country journeyed to the city. Then the white folks had to stay indoors, for every street was so packed with charging cavaliers and cavalieresses that it was next to impossible to thread one's way through the cavalcades without getting crippled.

At night they feasted and the girls danced the lascivious *hula hula* — a dance that is said to exhibit the very perfection of educated motion of limb and arm, hand, head, and body, and the exactest uniformity of movement and accuracy of "time." It was performed by a circle of girls with no raiment on them to speak of, who went through an infinite variety of motions and figures without prompting, and yet so true was their "time," and in such perfect concert did they move that when they were placed in a straight line, hands, arms, bodies, limbs, and heads waved, swayed, gesticulated, bowed, stooped, whirled, squirmed, twisted, and undulated as if they were part and parcel of a single individual; and it was difficult to believe they were not moved in a body by some exquisite piece of mechanism.

Of late years, however, Saturday has lost most of its quondam gala features. This weekly stampede of the natives interfered too much with labor and the interests of the white folks, and by sticking in a law here, and preaching a sermon there, and by various other means, they gradually broke it up.

The demoralizing *hula hula* was forbidden to be performed, save at night, with closed doors, in presence of few spectators, and only by permission duly procured from the authorities and the payment of ten dollars for the same. There are few girls now-a-days able to dance this ancient national dance in the highest perfection of the art.

The missionaries have christianized and educated all the natives. They all belong to the church, and there is not one of them, above the age of eight years, but can read and write with facility in the native tongue. It is the most universally educated race of people outside of China. They have any quantity of books, printed in the Kanaka language, and all the natives are fond of reading. They are inveterate church-goers — nothing can keep them away. All this ameliorating cultivation has at last built up in the native women a profound respect for chastity — in other people. Perhaps that is enough to say on that head. The national sin will die out when the race does, but perhaps not earlier. But doubtless this purifying is not far off, when we re-flect that contact with civilization and the whites has reduced the native population from *four hundred thousand* (Captain Cook's estimate), to *fifty-five thousand* in something over eighty years!

Society is a queer medley in this notable mission-ary, whaling, and governmental center. If you get into conversation with a stranger and experience that natural desire to know what sort of ground you are treading on by finding out what manner of man your stranger is, strike out boldly and address him

as "Captain." Watch him narrowly, and if you see
by his countenance that you are on the wrong tack,
ask him where he preaches. It is a safe bet that
he is either a missionary or captain of a whaler. I
am now personally acquainted with seventy-two captains and ninety-six missionaries. The captains and
ministers form one-half of the population; the third
fourth is composed of common Kanakas and mer-
cantile foreigners and their families, and the final
fourth is made up of high officers of the Hawaiian
government. And there are just about cats enough
for three apiece all around.

A solemn stranger met me in the suburbs the
other day, and said:

"Good morning, your reverence. Preach in the
stone church yonder, no doubt?"

"No, I don't. I'm not a preacher."

"Really, I beg your pardon, Captain. I trust
you had a good season. How much oil"—

"Oil? What do you take me for? I'm not a
whaler."

"Oh, I beg a thousand pardons, your Excellency.
Major-General in the household troops, no doubt?
Minister of the Interior, likely? Secretary of War?
First Gentleman of the Bed-chamber? Commis-
sioner of the Royal"—

"Stuff! I'm no official. I'm not connected in
any way with the government."

"Bless my life! Then who the mischief are you?
what the mischief are you? and how the mischief

did you get here, and where in thunder did you come from?"

"I'm only a private personage — an unassuming stranger — lately arrived from America."

"No? Not a missionary! Not a whaler! not a member of his Majesty's Government! not even Secretary of the Navy! Ah, Heaven! it is too blissful to be true; alas, I do but dream. And yet that noble, honest countenance — those oblique, ingenuous eyes — that massive head, incapable of— of — anything; your hand; give me your hand, bright waif. Excuse these tears. For sixteen weary years I have yearned for a moment like this, and "—

Here his feelings were too much for him, and he swooned away. I pitied this poor creature from the bottom of my heart. I was deeply moved. I shed a few tears on him and kissed him for his mother. I then took what small change he had and "shoved."

CHAPTER XXVI.

I STILL quote from my journal:
I found the national legislature to consist of half a dozen white men and some thirty or forty natives. It was a dark assemblage. The nobles and ministers (about a dozen of them altogether) occupied the extreme left of the hall, with David Kalakaua (the King's Chamberlain) and Prince William at the head. The President of the Assembly, His Royal Highness M. Kekuanaoa,* and the vice-president (the latter a white man) sat in the pulpit, if I may so term it.

The President is the King's father. He is an erect, strongly built, massive featured, white-haired, tawny old gentleman of eighty years of age or thereabouts. He was simply but well dressed, in a blue cloth coat and white vest, and white pantaloons, without spot, dust, or blemish upon them. He bears himself with a calm, stately dignity, and is a man of noble presence. He was a young man and a distinguished warrior under that terrific fighter, Kamehameha I., more than half a century ago. A

* Since dead.

knowledge of his career suggested some such thought as this: "This man, naked as the day he was born, and war-club and spear in hand, has charged at the head of a horde of savages against other hordes of savages more than a generation and a half ago, and reveled in slaughter and carnage; has worshiped wooden images on his devout knees; has seen hundreds of his race offered up in heathen temples as sacrifices to wooden idols, at a time when no missionary's foot had ever pressed this soil, and he had never heard of the white man's God; has believed his enemy could secretly pray him to death; has seen the day, in his childhood, when it was a crime punishable by death for a man to eat with his wife, or for a plebeian to let his shadow fall upon the king — and now look at him: an educated Christian; neatly and handsomely dressed; a high-minded, elegant gentleman; a traveler, in some degree, and one who has been the honored guest of royalty in Europe; a man practiced in holding the reins of an enlightened government, and well versed in the politics of his country and in general, practical information. Look at him, sitting there presiding over the deliberations of a legislative body, among whom are white men — a grave, dignified, statesmanlike personage, and as seemingly natural and fitted to the place as if he had been born in it and had never been out of it in his lifetime. How the experiences of this old man's eventful life shame the cheap inventions of romance!"

Kekuanaoa is not of the blood royal. He derives
his princely rank from his wife, who was a daughter
of Kamehameha the Great. Under other monarchies
the male line takes precedence of the female in
tracing genealogies, but here the opposite is the
case — the female line takes precedence. Their
reason for this is exceedingly sensible, and I recom-
mend it to the aristocracy of Europe: They say it
is easy to know who a man's mother was, but, etc.,
etc.

The christianizing of the natives has hardly even
weakened some of their barbarian superstitions,
much less destroyed them. I have just referred to
one of these. It is still a popular belief that if your
enemy can get hold of any article belonging to you
he can get down on his knees over it and *pray you
to death*. Therefore many a native gives up and
dies merely because he *imagines* that some enemy is
putting him through a course of damaging prayer.
This praying an individual to death seems absurd
enough at a first glance, but then when we call to
mind some of the pulpit efforts of certain of our
own ministers the thing looks plausible.

In former times, among the Islanders, not only a
plurality of wives was customary, but a *plurality of
husbands* likewise. Some native women of noble
rank had as many as six husbands. A woman thus
supplied did not reside with all her husbands at
once, but lived several months with each in turn.
An understood sign hung at her door during these

months. When the sign was taken down, it meant
" NEXT."

In those days woman was rigidly taught to " know
her place." Her place was to do all the work, take
all the cuffs, provide all the food, and content herself with what was left after her lord had finished his
dinner. She was not only forbidden, by ancient
law, and under penalty of death, to eat with her
husband or enter a canoe, but was debarred, under
the same penalty, from eating bananas, pineapples,
oranges, and other choice fruits at any time or in
any place. She had to confine herself pretty
strictly to " poi " and hard work. These poor ignorant heathen seem to have had a sort of groping
idea of what came of woman eating fruit in the
garden of Eden, and they did not choose to take
any more chances. But the missionaries broke up
this satisfactory arrangement of things. They
liberated woman and made her the equal of man.

The natives had a romantic fashion of burying
some of their children alive when the family became
larger than necessary. The missionaries interfered
in this matter too, and stopped it.

To this day the natives are able to *lie down and
die whenever they want to*, whether there is anything
the matter with them or not. If a Kanaka takes a
notion to die, that is the end of him; nobody can
persuade him to hold on; all the doctors in the
world could not save him.

A luxury which they enjoy more than anything

else, is a large funeral. If a person wants to get rid of a troublesome native, it is only necessary to promise him a fine funeral and name the hour, and he will be on hand to the minute — at least his remains will.

All the natives are Christians, now, but many of them still desert to the Great Shark God for temporary succor in time of trouble. An irruption of the great volcano of Kilauea, or an earthquake, always brings a deal of latent loyalty to the Great Shark God to the surface. It is common report that the king, educated, cultivated, and refined Christian gentleman as he undoubtedly is, still turns to the idols of his fathers for help when disaster threatens. A planter caught a shark, and one of his christianized natives testified his emancipation from the thrall of ancient superstition by assisting to dissect the shark after a fashion forbidden by his abandoned creed. But remorse shortly began to torture him. He grew moody and sought solitude; brooded over his sin, refused food, and finally said he must die and ought to die, for he had sinned against the Great Shark God and could never know peace any more. He was proof against persuasion and ridicule, and in the course of a day or two took to his bed and died, although he showed no symptom of disease. His young daughter followed his lead and suffered a like fate within the week. Superstition is ingrained in the native blood and bone and it is only natural that it should crop out in time of distress. Wherever one goes in the Islands, he will find small piles of stones by the wayside, covered with leafy offerings, placed there by the natives to appease evil spirits or honor local deities belonging to the mythology of former days.

In the rural districts of any of the Islands, the traveler hourly comes upon parties of dusky maidens bathing in the streams or in the sea without any clothing on and exhibiting no very intemperate zeal in the matter of hiding their nakedness. When the missionaries first took up their residence in Honolulu, the native women would pay their families frequent friendly visits, day by day, not even clothed with a blush. It was found a hard matter to convince them that this was rather indelicate. Finally, the missionaries provided them with long, loose calico robes, and that ended the difficulty — for the women would troop through the town, stark naked, with their robes folded under their arms, march to the missionary houses and then proceed to dress! The natives soon manifested a strong proclivity for clothing, but it was shortly apparent that they only wanted it for grandeur. The missionaries imported a quantity of hats, bonnets, and other male and female wearing apparel, instituted a general distribution, and begged the people not to come to church naked, next Sunday, as usual. And they did not; but the national spirit of unselfishness led them to divide up with neighbors who were not at the distribution, and next Sabbath the poor preachers could

16**

hardly keep countenance before their vast congrega-
tions. In the midst of the reading of a hymn a
brown, stately dame would sweep up the aisle with
a world of airs, with nothing in the world on but a
" stovepipe " hat and a pair of cheap gloves; another
dame would follow, tricked out in a man's shirt, and
nothing else; another one would enter with a flourish,
with simply the sleeves of a bright calico dress tied
around her waist and the rest of the garment dragging behind like a peacock's tail off duty; a stately
" buck " Kanaka would stalk in with a woman's
bonnet on, wrong side before — only this, and nothing more; after him would stride his fellow, with
the legs of a pair of pantaloons tied around his neck,
the rest of his person untrammeled; in his rear
would come another gentleman simply gotten up in
a fiery necktie and a striped vest. The poor creatures were beaming with complacency and wholly
unconscious of any absurdity in their appearance.
They gazed at each other with happy admiration,
and it was plain to see that the young girls were
taking note of what each other had on, as naturally
as if they had always lived in a land of Bibles and
knew what churches were made for; here was the
evidence of a dawning civilization. The spectacle
which the congregation presented was so extraordinary and withal so moving, that the missionaries
found it difficult to keep to the text and go on with
the services; and by and by when the simple children of the sun began a general swapping of gar-

ments in open meeting and produced some irresistibly grotesque effects in the course of re-dressing, there was nothing for it but to cut the thing short with the benediction and dismiss the fantastic assemblage.

In our country, children play "keep house"; and in the same high-sounding but miniature way the grown folk here, with the poor little material of slender territory and meager population, play "empire." There is his Royal Majesty, the King, with a New York detective's income of thirty or thirty-five thousand dollars a year from the "royal civil list" and the "royal domain." He lives in a two-story frame "palace."

And there is the "royal family"—the customary hive of royal brothers, sisters, cousins, and other noble drones and vagrants usual to monarchy,—all with a spoon in the national pap-dish, and all bearing such titles as his or her Royal Highness the Prince or Princess So-and-so. Few of them can carry their royal splendors far enough to ride in carriages, however; they sport the economical Kanaka horse or "hoof it"* with the plebeians.

Then there is his Excellency the "Royal Chamberlain"—a sinecure, for his Majesty dresses himself with his own hands, except when he is ruralizing at Waikiki, and then he requires no dressing.

Next we have his Excellency the Commander-in-chief of the Household Troops, whose forces consist

* Missionary phrase.

P**

of about the number of soldiers usually placed under a corporal in other lands.

Next comes the Royal Steward and the Grand Equerry in Waiting — high dignitaries with modest salaries and little to do.

Then we have his Excellency the First Gentleman of the Bed-chamber — an office as easy as it is magnificent.

Next we come to his Excellency the Prime Minister, a renegade American from New Hampshire, all jaw, vanity, bombast, and ignorance, a lawyer of "shyster" caliber, a fraud by nature, a humble worshiper of the scepter above him, a reptile never tired of sneering at the land of his birth or glorifying the ten-acre kingdom that has adopted him — salary, $4,000 a year, vast consequence, and no perquisites.

Then we have his Excellency the Imperial Minister of Finance, who handles a million dollars of public money a year, sends in his annual "budget" with great ceremony, talks prodigiously of "finance," suggests imposing schemes for paying off the "national debt" (of $150,000), and does it all for $4,000 a year and unimaginable glory.

Next we have his Excellency the Minister of War, who holds sway over the royal armies — they consist of two hundred and thirty uniformed Kanakas, mostly Brigadier-Generals, and if the country ever gets into trouble with a foreign power we shall probably hear from them. I knew an American whose copper-plate visiting card bore this impressive legend: "Lieutenant-Colonel in the Royal Infantry." To say that he was proud of this distinction is stating it but tamely. The Minister of War has also in his charge some venerable swivels on Punch-Bowl Hill wherewith royal salutes are fired when foreign vessels of war enter the port.

Next comes his Excellency the Minister of the Navy—a nabob who rules the "royal fleet" (a steam tug and a sixty-ton schooner).

And next comes his Grace the Lord Bishop of Honolulu, the chief dignitary of the "Established Church"—for when the American Presbyterian missionaries had completed the reduction of the nation to a compact condition of Christianity, native royalty stepped in and erected the grand dignity of an "Established (Episcopal) Church" over it, and imported a cheap ready-made bishop from England to take charge. The chagrin of the missionaries has never been comprehensively expressed, to this day, profanity not being admissible.

Next comes his Excellency the Minister of Public Instruction.

Next, their Excellencies the Governors of Oahu, Hawaii, etc., and after them a string of High Sheriffs and other small fry too numerous for computation.

Then there are their Exellencies the Envoy Extraordinary and Minister Plenipotentiary of his Imperial Majesty the Emperor of the French; her British Majesty's Minister; the Minister Resident of the

United States; and some six or eight representatives of other foreign nations, all with sounding titles, imposing dignity, and prodigious but economical state.

Imagine all this grandeur in a playhouse "kingdom" whose population falls absolutely short of sixty thousand souls!

The people are so accustomed to nine-jointed titles and colossal magnates that a foreign prince makes very little more stir in Honolulu than a Western Congressman does in New York.

And let it be borne in mind that there is a strictly defined "court costume" of so "stunning" a nature that it would make the clown in a circus look tame and commonplace by comparison; and each Hawaiian official dignitary has a gorgeous varicolored, gold-laced uniform peculiar to his office — no two of them are alike, and it is hard to tell which one is the "loudest." The king has a "drawing-room" at stated intervals, like other monarchs, and when these varied uniforms congregate there weak-eyed people have to contemplate the spectacle through smoked glass. Is there not a gratifying contrast between this latter-day exhibition and the one the ancestors of some of these magnates afforded the missionaries the Sunday after the old-time distribution of clothing? Behold what religion and civilization have wrought!

CHAPTER XXVII.

WHILE I was in Honolulu I witnessed the ceremonious funeral of the King's sister, her Royal
Highness the Princess Victoria. According to the
royal custom, the remains had lain in state at the
palace *thirty days*, watched day and night by a
guard of honor. And during all that time a great
multitude of natives from the several islands had
kept the palace grounds well crowded and had made
the place a pandemonium every night with their
howlings and wailings, beating of tom-toms and
dancing of the (at other times) forbidden " hula-
hula " by half-clad maidens to the music of songs of
questionable decency chanted in honor of the deceased. The printed program of the funeral procession interested me at the time; and after what
I have just said of Hawaiian grandiloquence in the
matter of " playing empire," I am persuaded that a
perusal of it may interest the reader:

After reading the long list of dignitaries, etc., and remembering the
sparseness of the population, one is almost inclined to wonder where the
material for that portion of the procession devoted to " Hawaiian Population Generally " is going to be procured:

Undertaker.

Royal School. Kawaiahao School. Roman Catholic School. Miæmæ
School.

Honolulu Fire Department.

Mechanics' Benefit Union.

Attending Physicians.

Konohikis (Superintendents) of the Crown Lands, Konohikis of the
Private Lands of His Majesty, Konohikis of Private Lands
of Her late Royal Highness.

Governor of Oahu and Staff.

Hulumanu (Military Company).

Household Troops.

The Prince of Hawaii's Own (Military Company).

The King's household servants.

Servants of Her late Royal Highness.

Protestant Clergy. The Clergy of the Roman Catholic Church.

His Lordship Louis Maigret, The Right Rev. Bishop of Arathea, Vicar-
Apostolic of the Hawaiian Islands.

The Clergy of the Hawaiian Reformed Catholic Church.

His Lordship the Right Rev. Bishop of Honolulu.

Escort Hawaiian Cavalry.
Large Kahilis.
Small Kahilis.
Pall Bearers.

[HEARSE.]

Escort Hawaiian Cavalry.
Large Kahilis. *
Small Kahilis.
Pall Bearers.

Her Majesty Queen Emma's Carriage.

His Majesty's Staff.

Carriage of Her late Royal Highness.

Carriage of Her Majesty the Queen Dowager.

The King's Chancellor.

Cabinet Ministers.

His Excellency the Minister Resident of the United States.

---

* Ranks of long-handled mops made of gaudy feathers — sacred to
royalty. They are stuck in the ground around the tomb and left there

H. I. M's Commissioner.
H. B. M's Acting Commissioner.
Judges of Supreme Court.
Privy Councillors.
Members of Legislative Assembly.
Consular Corps.
Circuit Judges.
Clerks of Government Departments.
Members of the Bar.
Collector General, Custom-house Officers and Officers of the Customs.
Marshal and Sheriffs of the different Islands.
King's Yeomanry.
Foreign Residents.
Ahahui Kaahumanu.
Hawaiian Population Generally.
Hawaiian Cavalry.
Police Force.

I resume my journal at the point where the procession arrived at the royal mausoleum:

As the procession filed through the gate the military deployed handsomely to the right and left and formed an avenue through which the long column of mourners passed to the tomb. The coffin was borne through the door of the mausoleum, followed by the King and his chiefs, the great officers of the kingdom, foreign Consuls, Embassadors, and distinguished guests (Burlingame and General Van Valkenburgh). Several of the kahilis were then fastened to a frame-work in front of the tomb, there to remain until they decay and fall to pieces, or, forestalling this, until another scion of royalties dies. At this point of the proceedings the multitude set up such a heart-broken wailing as I hope never to hear again. The soldiers fired three volleys of musketry—the wailing being previously silenced to permit of the guns being heard. His Highness Prince William, in a showy military uniform (the "true prince," this—scion of the house overthrown by the present dynasty —he was formerly betrothed to the Princess but was not allowed to marry her), stood guard and paced back and forth within the door. The privileged few who followed the coffin into the mausoleum remained some time, but the King soon came out and stood in the door and near one side of it. A stranger could have guessed his rank (although he was so simply and unpretentiously dressed) by the profound deference paid him by all persons in his vicinity; by seeing his high officers receive his quiet orders and suggestions with bowed and uncovered heads; and by observing how careful those persons who came out of the mausoleum were to avoid " crowding " him (although there was room enough in the doorway for a wagon to pass, for that matter); how respectfully they edged out sideways, scraping their backs against the wall and always presenting a front view of their persons to his Majesty, and never putting their hats on until they were well out of the royal presence.

He was dressed entirely in black — dress-coat and silk hat — and looked rather democratic in the midst of the showy uniforms about him. On his breast he wore a large gold star, which was half hidden by the lapel of his coat. He remained at the door a half hour, and occasionally gave an order to the men who were erecting the *kahilis* before the tomb. He had the good taste to make one of them substitute black crape for the ordinary hempen rope he was about to tie one of them to the frame-work with. Finally he entered his carriage and drove away, and the populace shortly began to drop into his wake. While he was in view there was but one man who attracted more attention than himself, and that was Harris (the Yankee Prime Minister). This feeble personage had crape enough around his hat to express the grief of an entire nation, and as usual he neglected no opportunity of making himself conspicuous and exciting the admiration of the simple Kanakas. Oh! noble ambition of this modern Richelieu!

It is interesting to contrast the funeral ceremonies of the Princess Victoria with those of her noted ancestor Kamehameha the Conqueror, who died fifty years ago — in 1819, the year before the first missionaries came.

"On the 8th of May, 1819, at the age of sixty-six, he died, as he had lived, in the faith of his country. It was his misfortune not to have come in contact with men who could have rightly influenced his religious aspirations. Judged by his advantages and compared with the most eminent of his countrymen he may be justly styled not only great, but good. To this day his memory warms the heart and elevates the national feelings of Hawaiians. They are proud of their old warrior King; they love his name; his deeds form their historical age; and an enthusiasm everywhere prevails, shared even by foreigners who knew his worth, that constitutes the firmest pillar of the throne of his dynasty.

"In lieu of human victims (the custom of that age), a sacrifice of three hundred dogs attended his obsequies — no mean holocaust when their national value and the estimation in which they were held are considered. The bones of Kamehameha, after being kept for a while, were so carefully concealed that all knowledge of their final resting place is now lost. There was a proverb current among the common people that the bones of a cruel King could not be hid; they made fish-hooks and arrows of them, upon which, in using them, they vented their abhorrence of his memory in bitter execrations."

The account of the circumstances of his death, as written by the native historians, is full of minute detail, but there is scarcely a line of it which does not mention or illustrate some bygone custom of the country. In this respect it is the most comprehensive document I have yet met with. I will quote it entire:

"When Kamehameha was dangerously sick, and the priests were unable to cure him, they said: 'Be of good courage and build a house for the god (his own private god or idol), that thou mayest recover.' The chiefs corroborated this advice of the priests, and a place of worship was prepared for Kukailimoku, and consecrated in the evening. They proposed also to the King, with a view to prolong his life, that human victims should be sacrificed to his deity; upon which the greater part of the people absconded through fear of death, and concealed themselves in hiding places till the *tabu** in which destruction impended, was past. It is doubtful whether Kamehameha approved of the plan of the chiefs and priests to sacrifice men, as he was known to say, 'The men are sacred for the King;' meaning that they were for the service

---

* *Tabu* (pronounced tah-boo), means prohibition (we have borrowed it), or sacred. The tabu was sometimes permanent, sometimes temporary; and the person or thing placed under tabu was for the time being sacred to the purpose for which it was set apart. In the above case the victims selected under the tabu would be sacred to the sacrifice.

of his successor. This information was derived from Liholiho, his son.

"After this, his sickness increased to such a degree that he had not strength to turn himself in his bed. When another season, consecrated for worship at the new temple (*heiau*) arrived, he said to his son, Liholiho, 'Go thou and make supplication to thy god; I am not able to go, and will offer my prayers at home.' When his devotions to his feathered god, Kukailimoku, were concluded, a certain religiously disposed individual, who had a bird god, suggested to the King that through its influence his sickness might be removed. The name of this god was Pua; its body was made of a bird, now eaten by the Hawaiians, and called in their language *alae*. Kamehameha was willing that a trial should be made, and two houses were constructed to facilitate the experiment; but while dwelling in them he became so very weak as not to receive food. After lying there three days, his wives, children, and chiefs, perceiving that he was very low, returned him to his own house. In the evening he was carried to the eating house,* where he took a little food in his mouth which he did not swallow; also a cup of water. The chiefs requested him to give them his counsel; but he made no reply, and was carried back to the dwelling house; but when near midnight — ten o'clock, perhaps — he was carried again to the place to eat; but, as before, he merely tasted of what was presented to him. Then Kaikioewa addressed him thus: 'Here we all are, your younger brethren, your son Liholiho and your foreigner; impart to us your dying charge, that Liholiho and Kaahumanu may hear.' Then Kamehameha inquired, 'What do you say?' Kaikioewa repeated, 'Your counsels for us.' He then said, 'Move on in my good way and—.' He could proceed no further. The foreigner, Mr. Young, embraced and kissed him. Hoapili also embraced him, whispering something in his ear, after which he was taken back to the house. About twelve he was carried once more to the house for eating, into which his head entered, while his body was in the dwelling house immediately adjoining. It should be remarked that this frequent carrying of a sick chief from one house to another resulted from the *tabu* system, then in force. There were at that time six houses (huts) connected with an establishment — one was for worship, one for the men

---

* It was deemed pollution to eat in the same hut a person slept in — the fact that the patient was dying could not modify the rigid etiquette.

to eat in, an eating house for the women, a house to sleep in, a house in which to manufacture kapa (native cloth), and one where, at certain intervals, the women might dwell in seclusion.

"The sick was once more taken to his house, when he expired; this was at two o'clock, a circumstance from which Leleiohoku derived his name. As he breathed his last, Kalaimoku came to the eating house to order those in it to go out. There were two aged persons thus directed to depart; one went, the other remained on account of love to the King, by whom he had formerly been kindly sustained. The children also were sent away. Then Kalaimoku came to the house, and the chiefs had a consultation. One of them spoke thus: 'This is my thought — we will eat him raw.'* Kaahumanu (one of the dead King's widows) replied, ' Perhaps his body is not at our disposal; that is more properly with his successor. Our part in him — his breath — has departed; his remains will be disposed of by Liholiho.'

"After this conversation the body was taken into the consecrated house for the performance of the proper rites by the priest and the new King. The name of this ceremony is *uko;* and when the sacred hog was baked the priest offered it to the dead body, and it became a god, the King at the same time repeating the customary prayers.

"Then the priest, addressing himself to the King and chiefs, said: 'I will now make known to you the rules to be observed respecting persons to be sacrificed on the burial of this body. If you obtain one man before the corpse is removed, one will be sufficient; but after it leaves this house four will be required. If delayed until we carry the corpse to the grave there must be ten; but after it is deposited in the grave there must be fifteen. To-morrow morning there will be a *tabu*, and, if the sacrifice be delayed until that time, forty men must die.'

"Then the high priest, Hewahewa, inquired of the chiefs, 'Where shall be the residence of King Liholiho?' They replied, 'Where, indeed? You, of all men, ought to know.' Then the priest observed, 'There are two suitable places; one is Kau, the other is Kohala.' The chiefs preferred the latter, as it was more thickly inhabited. The priest

---

*This sounds suspicious, in view of the fact that all Sandwich Island historians, white and black, protest that cannibalism never existed in the islands. However, since they only proposed to " eat him raw " we "won't count that." But it would certainly have been cannibalism if they had cooked him.—[M. T.]

added, 'These are proper places for the King's residence; but he must not remain in Kona, for it is polluted.' This was agreed to. It was now break of day. As he was being carried to the place of burial the people perceived that their King was dead, and they wailed. When the corpse was removed from the house to the tomb, a distance of one chain, the procession was met by a certain man who was ardently attached to the deceased. He leaped upon the chiefs who were carry- ing the King's body; he desired to die with him on account of his love. The chiefs drove him away. He persisted in making numerous attempts, which were unavailing. Kalaimoku also had it in his heart to die with him, but was prevented by Hookio.

"The morning following Kamehameha's death, Liholiho and his train departed for Kohala, according to the suggestions of the priest, to avoid the defilement occasioned by the dead. At this time if a chief died the land was polluted, and the heirs sought a residence in another part of the country until the corpse was dissected and the bones tied in a bundle, which being done, the season of defilement terminated. If the deceased were not a chief, the house only was defiled, which became pure again on the burial of the body. Such were the laws on this subject.

"On the morning on which Liholiho sailed in his canoe for Kohala, the chiefs and people mourned after their manner on occasion of a chief's death, conducting themselves like madmen and like beasts. Their conduct was such as to forbid description. The priests, also, put into action the sorcery apparatus, that the person who had prayed the King to death might die; for it was not believed that Kamehameha's departure was the effect either of sickness or old age. When the sorcerers set up by their fireplaces a stick with a strip of kapa flying at the top, the chief Keeaumoku, Kaahumanu's brother, came in a state of intoxication and broke the flagstaff of the sorcerers, from which it was inferred that Kaahumanu and her friends had been instrumental in the King's death. On this account they were subjected to abuse."

You have the contrast, now, and a strange one it is. This great queen, Kaahumanu, who was "sub- jected to abuse" during the frightful orgies that followed the king's death, in accordance with ancient custom, afterward became a devout Christian and a steadfast and powerful friend of the missionaries.

Dogs were, and still are, reared and fattened for food, by the natives — hence the reference to their value in one of the above paragraphs.

Forty years ago it was the custom in the Islands to suspend all law for a certain number of days after the death of a royal personage; and then a saturnalia ensued which one may picture to himself after a fashion, but not in the full horror of the reality. The people shaved their heads, knocked out a tooth or two, plucked out an eye sometimes, cut, bruised, mutilated or burned their flesh, got drunk, burned each other's huts, maimed or murdered one another according to the caprice of the moment, and both sexes gave themselves up to brutal and unbridled licentiousness. And after it all, came a torpor from which the nation slowly emerged bewildered and dazed, as if from a hideous half-remembered nightmare. They were not the salt of the earth, those " gentle children of the sun."

The natives still keep up an old custom of theirs which cannot be comforting to an invalid. When they think a sick friend is going to die, a couple of dozen neighbors surround his hut and keep up a deafening wailing night and day till he either dies or gets well. No doubt this arrangement has helped many a subject to a shroud before his appointed time.

They surround a hut and wail in the same heart-broken way when its occupant returns from a journey. This is their dismal idea of a welcome. A very little of it would go a great way with most of us.

CHAPTER XXVIII.

BOUND for Hawaii (a hundred and fifty miles distant), to visit the great volcano and behold the other notable things which distinguish that island above the remainder of the group, we sailed from Honolulu on a certain Saturday afternoon, in the good schooner *Boomerang*.

The *Boomerang* was about as long as two street cars, and about as wide as one. She was so small (though she was larger than the majority of the inter-island coasters) that when I stood on her deck I felt but little smaller than the Colossus of Rhodes must have felt when he had a man-of-war under him. I could reach the water when she lay over under a strong breeze. When the captain and my comrade (a Mr. Billings), myself and four other persons were all assembled on the little after portion of the deck which is sacred to the cabin passengers, it was full — there was not room for any more quality folks. Another section of the deck, twice as large as ours, was full of natives of both sexes, with their customary dogs, mats, blankets, pipes, calabashes of poi, fleas, and other luxuries and bag-

gage of minor importance. As soon as we set sail
the natives all lay down on the deck as thick as
negroes in a slave-pen, and smoked, conversed, and
spit on each other, and were truly sociable.

The little low-ceiled cabin below was rather larger
than a hearse, and as dark as a vault. It had two
coffins on each side — I mean two bunks. A small
table, capable of accommodating three persons at
dinner, stood against the forward bulkhead, and
over it hung the dingiest whale oil lantern that ever
peopled the obscurity of a dungeon with ghostly
shapes. The floor room unoccupied was not extensive. One might swing a cat in it, perhaps, but
not a long cat. The hold forward of the bulkhead
had but little freight in it, and from morning till
night a portly old rooster, with a voice like Baalam's
ass, and the same disposition to use it, strutted up
and down in that part of the vessel and crowed. He
usually took dinner at six o'clock, and then, after
an hour devoted to meditation, he mounted a barrel
and crowed a good part of the night. He got
hoarser and hoarser all the time, but he scorned to
allow any personal consideration to interfere with
his duty, and kept up his labors in defiance of
threatened diphtheria.

Sleeping was out of the question when he was on
watch. He was a source of genuine aggravation
and annoyance. It was worse than useless to shout
at him or apply offensive epithets to him — he only
took these things for applause, and strained himself

17**

to make more noise. Occasionally, during the day, I threw potatoes at him through an aperture in the bulkhead, but he only dodged and went on crowing.

The first night, as I lay in my coffin, idly watching the dim lamp swinging to the rolling of the ship, and snuffing the nauseous odors of bilge water, I felt something gallop over me. I turned out promptly. However, I turned in again when I found it was only a rat. Presently something galloped over me once more. I knew it was not a rat this time, and I thought it might be a centipede, because the captain had killed one on deck in the afternoon. I turned out. The first glance at the pillow showed me a repulsive sentinel perched upon each end of it — cockroaches as large as peach leaves — fellows with long, quivering antennæ and fiery, malignant eyes. They were grating their teeth like tobacco worms, and appeared to be dissatisfied about something. I had often heard that these reptiles were in the habit of eating off sleeping sailors' toe nails down to the quick, and I would not get in the bunk any more. I lay down on the floor. But a rat came and bothered me, and shortly afterward a procession of cockroaches arrived and camped in my hair. In a few moments the rooster was crowing with uncommon spirit, and a party of fleas were throwing double somersaults about my person in the wildest disorder, and taking a bite every time they struck. I was beginning to feel really annoyed. I got up and put my clothes on and went on deck.

The above is not overdrawn; it is a truthful sketch of inter-island schooner life. There is no such thing as keeping a vessel in elegant condition, when she carries molasses and Kanakas.

It was compensation for my sufferings to come unexpectedly upon so beautiful a scene as met my eye — to step suddenly out of the sepulchral gloom of the cabin and stand under the strong light of the moon — in the center, as it were, of a glittering sea of liquid silver — to see the broad sails straining in the gale, the ship keeled over on her side, the angry foam hissing past her lee bulwarks, and sparkling sheets of spray dashing high over her bows and raining upon her decks; to brace myself and hang fast to the first object that presented itself, with hat jammed down and coat-tails whipping in the breeze, and feel that exhilaration that thrills in one's hair and quivers down his backbone when he knows that every inch of canvas is drawing and the vessel cleaving through the waves at her utmost speed. There was no darkness, no dimness, no obscurity there. All was brightness, every object was vividly defined. Every prostrate Kanaka; every coil of rope; every calabash of poi; every puppy; every seam in the flooring; every bolthead; every object, however minute, showed sharp and distinct in its every outline; and the shadow of the broad mainsail lay black as a pall upon the deck, leaving Billings's white upturned face glorified and his body in a total eclipse.

Monday morning we were close to the island of Hawaii. Two of its high mountains were in view — Mauna Loa and Hualaiai. The latter is an imposing peak, but being only ten thousand feet high is seldom mentioned or heard of. Mauna Loa is said to be sixteen thousand feet high. The rays of glittering snow and ice, that clasped its summit like a claw, looked refreshing when viewed from the blistering climate we were in. One could stand on that mountain (wrapped up in blankets and furs to keep warm), and while he nibbled a snowball or an icicle to quench his thirst he could look down the long sweep of its sides and see spots where plants are growing that grow only where the bitter cold of winter prevails; lower down he could see sections devoted to productions that thrive in the temperate zone alone; and at the bottom of the mountain he could see the home of the tufted cocoa-palms and other species of vegetation that grow only in the sultry atmosphere of eternal summer. He could see all the climes of the world at a single glance of the eye, and that glance would only pass over a distance of four or five miles as the bird flies!

By and by we took boat and went ashore at Kailua, designing to ride horseback through the pleasant orange and coffee region of Kona, and rejoin the vessel at a point some leagues distant. This journey is well worth taking. The trail passes

along on high ground — say a thousand feet above
sea level — and usually about a mile distant from the
ocean, which is always in sight, save that occasionally you find yourself buried in the forest in the
midst of a rank tropical vegetation and a dense
growth of trees, whose great boughs overarch the road
and shut out sun and sea and everything, and leave
you in a dim, shady tunnel, haunted with invisible
singing birds and fragrant with the odor of flowers.
It was pleasant to ride occasionally in the warm sun,
and feast the eye upon the ever-changing panorama
of the forest (beyond and below us), with its many
tints, its softened lights and shadows, its billowy
undulations sweeping gently down from the mountain to the sea. It was pleasant also, at intervals,
to leave the sultry sun and pass into the cool, green
depths of this forest and indulge in sentimental re-
flections under the inspiration of its brooding twilight
and its whispering foliage.

We rode through one orange grove that had ten
thousand trees in it! They were all laden with fruit.

At one farmhouse we got some large peaches of
excellent flavor. This fruit, as a general thing, does
not do well in the Sandwich Islands. It takes a sort
of almond shape, and is small and bitter. It needs
frost, they say, and perhaps it does; if this be so, it
will have a good opportunity to go on needing it, as
it will not be likely to get it. The trees from which
the fine fruit I have spoken of came had been
planted and replanted *sixteen times*, and to this

treatment the proprietor of the orchard attributed his success.

We passed several sugar plantations — new ones and not very extensive. The crops were, in most cases, third rattoons. [NOTE. The first crop is called " plant cane "; subsequent crops which spring from the original roots, without replanting, are called " rattoons."] Almost everywhere on the island of Hawaii sugar-cane matures in twelve months, both rattoons and plant, and although it ought to be taken off as soon as it tassels, no doubt, it is not absolutely necessary to do it until about four months afterward. In Kona, the average yield of an acre of ground is *two tons* of sugar, they say. This is only a moderate yield for these islands, but would be astounding for Louisiana and most other sugar-growing countries. The plantations in Kona being on pretty high ground — up among the light and frequent rains — no irrigation whatever is required.

CHAPTER XXIX.

WE stopped some time at one of the plantations, to rest ourselves and refresh the horses. We had a chatty conversation with several gentlemen present; but there was one person, a middle-aged man, with an absent look in his face, who simply glanced up, gave us good-day and lapsed again into the meditations which our coming had interrupted. The planters whispered us not to mind him — crazy. They said he was in the Islands for his health; was a preacher; his home, Michigan. They said that if he woke up presently and fell to talking about a correspondence which he had some time held with Mr. Greeley about a trifle of some kind, we must humor him and listen with interest; and we must humor his fancy that this correspondence was the talk of the world.

It was easy to see that he was a gentle creature and that his madness had nothing vicious in it. He looked pale, and a little worn, as if with perplexing thought and anxiety of mind. He sat a long time, looking at the floor, and at intervals muttering to himself and nodding his head acquiescingly or shaking it in mild protest. He was lost in his thought, or in his memories. We continued our talk with the planters, branching from subject to subject. But at last the word "circumstance," casually dropped, in the course of conversation, attracted his attention and brought an eager look into his countenance. He faced about in his chair and said:

"Circumstance? What circumstance? Ah, I know — I know too well. So you have heard of it too." [With a sigh.] "Well, no matter — all the world has heard of it. All the world. The whole world. It is a large world, too, for a thing to travel so far in — now, isn't it? Yes, yes — the Greeley correspondence with Erickson has created the saddest and bitterest controversy on both sides of the ocean — and still they keep it up! It makes us famous, but at what a sorrowful sacrifice! I was so sorry when I heard that it had caused that bloody and distressful war over there in Italy. It was little comfort to me, after so much bloodshed, to know that the victors sided with me, and the vanquished with Greeley. It is little comfort to know that Horace Greeley is responsible for the battle of Sadowa, and not me. Queen Victoria wrote me that she felt just as I did about it — she said that as much as she was opposed to Greeley and the spirit he showed in the correspondence with me, she would not have had Sadowa happen for hundreds of dollars. I can show you her letter, if you would like to see it. But, gentlemen, much as you may think you know about that unhappy correspondence, you cannot know the *straight* of it till you hear it from my lips. It has always been garbled in the journals, and even in history. Yes, even in history — think of it! Let me — *please* let me, give you the matter, exactly as it occurred. I truly will not abuse your confidence."

Then he leaned forward, all interest, all earnestness, and told his story — and told it appealingly, too, and yet in the simplest and most unpretentious way; indeed, in such a way as to suggest to one, all the time, that this was a faithful, honorable witness, giving evidence in the sacred interest of justice, and under oath. He said:

"Mrs. Beazeley — Mrs. Jackson Beazeley, widow, of the village of Campbellton, Kansas,— wrote me about a matter which was near her heart — a matter which many might think trivial, but to her it was a thing of deep concern. I was living in Michigan, then — serving in the ministry. She was, and is, an estimable woman — a woman to whom poverty and hardship have proven incentives to industry, in place of discouragements. Her only treasure was her son William, a youth just verging upon manhood; religious, amiable, and sincerely attached to agriculture. He was the widow's comfort and her pride. And so, moved by her love for him, she wrote me about a matter, as I have said before, which lay near her heart — because it lay near her boy's. She desired me to confer with Mr. Greeley about turnips. Turnips were the dream of her child's young ambition. While other youths were frittering away in frivolous amusements the precious years of budding vigor which God had given them for useful preparation, this boy was patiently enriching his mind with information concerning turnips. The sentiment which he felt toward the turnip was akin to adoration. He could not think of the turnip without emotion; he could not speak of it calmly; he could not contemplate it without exaltation; he could not eat it without shedding tears. All the poetry in his sensitive nature was in sympathy with the gracious vegetable. With the earliest pipe of dawn he sought his patch, and when the curtaining night drove him from it he shut himself up with his books and garnered statistics till sleep overcame him. On rainy days he sat and talked hours together with his mother about turnips. When company came, he made it his loving duty to put aside everything else and converse with them all the day long of his great joy in the turnip. And yet, was this joy rounded and complete? Was there no secret alloy of unhappiness in it? Alas, there was. There was a canker gnawing at his heart; the noblest inspiration of his soul eluded his endeavor — viz., he could not make of the turnip a climbing vine. Months went by; the bloom forsook his cheek, the fire faded out of his eye; sighings and abstraction usurped the place of smiles and cheerful converse.

But a watchful eye noted these things, and in time a motherly sympathy unsealed the secret. Hence the letter to me. She pleaded for attention — she said her boy was dying by inches.

"I was a stranger to Mr. Greeley, but what of that? The matter was urgent. I wrote and begged him to solve the difficult problem if possible, and save the student's life. My interest grew, until it partook of the anxiety of the mother. I waited in much suspense. At last the answer came.

"I found that I could not read it readily, the handwriting being unfamiliar and my emotions somewhat wrought up. It seemed to refer in part to the boy's case, but chiefly to other and irrelevant matters — such as paving-stones, electricity, oysters, and something which I took to be 'absolution' or 'agrarianism,' I could not be certain which; still, these appeared to be simply casual mentions, nothing more; friendly in spirit, without doubt, but lacking the connection or coherence necessary to make them useful. I judged that my understanding was affected by my feelings, and so laid the letter away till morning.

"In the morning I read it again, but with difficulty and uncertainty still, for I had lost some little rest and my mental vision seemed clouded. The note was more connected, now, but did not meet the emergency it was expected to meet. It was too discursive. It appeared to read as follows, though I was not certain of some of the words;

> 'Polygamy dissembles majesty; extracts redeem polarity; causes hitherto exist. Ovations pursue wisdom, or warts inherit and condemn. Boston, botany, cakes, folony undertakes, but who shall allay? We fear not. Yrxwly,   HEVACE EVEELOJ.'

"But there did not seem to be a word about turnips. There seemed to be no suggestion as to how they might be made to grow like vines. There was not even a reference to the Beazeleys. I slept upon the matter; I ate no supper, neither any breakfast next morning. So I resumed my work with a brain refreshed, and was very hopeful. *Now* the letter took a different aspect — all save the signature, which latter I judged to be only a harmless affectation of Hebrew. The epistle was necessarily from Mr. Greeley, for it bore the printed heading of *The Tribune*, and I had written to no one else there. The letter, I say, had taken a different aspect, but still its language was eccentric and avoided the issue. It now appeared to say:

> 'Bolivia extemporizes mackerel; borax esteems polygamy; sausages wither in the east. Creation perdu, is done; for woes inherent one can damn. Buttons, buttons, corks, geology underrates but we shall allay. My beer's out. Yrxwly,   HEVACE EVEELOJ.'

"I was evidently overworked. My comprehension was impaired. Therefore I gave two days to recreation, and then returned to my task greatly refreshed. The letter now took this form:

> 'Poultices do sometimes choke swine; tulips reduce posterity; causes leather to resist. Our notions empower wisdom, her let's afford while we can. Butter but any cakes, fill any undertaker, we'll wean him from his filly. We feel hot. Yrxwly,   HEVACE EVEELOJ.'

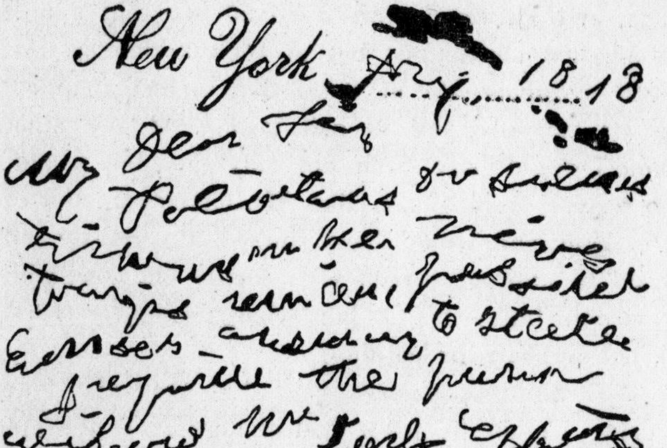

New York _____ 18_18

Dear Sir

yours,
Horace Greeley

"I was still not satisfied. These generalities did not meet the question. They were crisp, and vigorous, and delivered with a confidence that almost compelled conviction; but at such a time as this, with a human life at stake, they seemed inappropriate, worldly, and in bad taste. At any other time I would have been not only glad, but proud, to receive from a man like Mr. Greeley a letter of this kind, and would have studied it earnestly and tried to improve myself all I could; but now, with that poor boy in his far home languishing for relief, I had no heart for learning.

"Three days passed by, and I read the note again. Again its tenor had changed. It now appeared to say:

'Potations do sometimes wake wines; turnips restrain passion; causes necessary to state. Infest the poor widow; her lord's effects will be void. But dirt, bathing, etc., etc., followed unfairly, will worm him from his folly — so swear not. Yrxwly,   HEVACE EVEELOJ.'

"This was more like it. But I was unable to proceed. I was too much worn. The word 'turnips' brought temporary joy and encouragement, but my strength was so much impaired, and the delay might be so perilous for the boy, that I relinquished the idea of pursuing the translation further, and resolved to do what I ought to have done at first. I sat down and wrote Mr. Greeley as follows:

"DEAR SIR: I fear I do not entirely comprehend your kind note. It cannot be possible, Sir, that 'turnips restrain passion' — at least the

study or contemplation of turnips cannot — for it is this very employment that has scorched our poor friend's mind and sapped his bodily
strength. — But if they *do* restrain it, will you bear with us a little
further and explain how they should be prepared? I observe that you
say ' causes necessary to state,' but you have omitted to state them.

"Under a misapprehension, you seem to attribute to me interested
motives in this matter — to call it by no harsher term. But I assure
you, dear sir, that if I seem to be ' infesting the widow,' it is all *seeming*, and void of reality. It is from no seeking of mine that I am in
this position. She asked me, herself, to write you. I never have
infested her — indeed I scarcely know her. I do not infest anybody.
I try to go along, in my humble way, doing as near right as I can,
never harming anybody, and never *throwing out insinuations*. As for
' her lord and his effects,' they are of no interest to me. I trust I have
effects enough of my own — shall endeavor to get along with them, at
any rate, and not go mousing around to get hold of somebody's that
are ' void.' But do you not see? — this woman is a *widow* — she has
no ' lord.' He is dead — or pretended to be, when they buried him.
Therefore, no amount of ' dirt, bathing, etc., etc.,' howsoever ' unfairly
followed ' will be likely to ' worm him from his folly '— if being dead
and a ghost is ' folly.' Your closing remark is as unkind as it was
uncalled for; and if report says true you might have applied it to yourself, sir, with more point and less impropriety.

Very Truly Yours,   SIMON ERICKSON.

" In the course of a few days, Mr. Greeley did
what would have saved a world of trouble, and much
mental and bodily suffering and misunderstanding,
if he had done it sooner. To wit, he sent an intelli-
gible rescript or translation of his original note,
made in a plain hand by his clerk. Then the mystery cleared, and I saw that his heart had been
right, all the time. I will recite the note in its
clarified form:

[Translation.]

' Potatoes do sometimes make vines; turnips remain passive: cause
unnecessary to state. Inform the poor widow her lad's efforts will be

18**

vain. But diet, bathing, etc., etc., followed uniformly, will wean him from his folly — so fear not.   Yours,   HORACE GREELEY.'

"But alas, it was too late, gentlemen — too late. The criminal delay had done its work — young Beazeley was no more. His spirit had taken its flight to a land where all anxieties shall be charmed away, all desires gratified, all ambitions realized. Poor lad, they laid him to his rest with a turnip in each hand."

So ended Erickson, and lapsed again into nodding, mumbling, and abstraction. The company broke up, and left him so. . . . But they did not say what drove him crazy. In the momentary confusion, I forgot to ask.

CHAPTER XXX.

AT four o'clock in the afternoon we were winding down a mountain of dreary and desolate lava to the sea, and closing our pleasant land journey. This lava is the accumulation of ages; one torrent of fire after another has rolled down here in old times, and built up the island structure higher and higher. Underneath, it is honeycombed with caves. It would be of no use to dig wells in such a place; they would not hold water — you would not find any for them to hold, for that matter. Consequently, the planters depend upon cisterns.

The last lava flow occurred here so long ago that there are none now living who witnessed it. In one place it enclosed and burned down a grove of cocoa-nut trees, and the holes in the lava where the trunks stood are still visible: their sides retain the impression of the bark: the trees fell upon the burning river, and becoming partly submerged, left in it the perfect counterpart of every knot and branch and leaf, and even nut, for curiosity-seekers of a long distant day to gaze upon and wonder at.

There were doubtless plenty of Kanaka sentinels

on guard hereabouts at that time, but they did not leave casts of their figures in the lava as the Roman sentinels at Herculaneum and Pompeii did. It is a pity it is so, because such things are so interesting; but so it is. They probably went away. They went away early, perhaps. However, they had their merits; the Romans exhibited the higher pluck, but the Kanakas showed the sounder judgment.

Shortly, we came in sight of that spot whose history is so familiar to every schoolboy in the wide world — Kealakekua Bay — the place where Captain Cook, the great circumnavigator, was killed by the natives, nearly a hundred years ago. The setting sun was flaming upon it, a summer shower was falling, and it was spanned by two magnificent rainbows. Two men who were in advance of us rode through one of these and for a moment their garments shone with a more than regal splendor. Why did not Captain Cook have taste enough to call his great discovery the Rainbow Islands? These charming spectacles are present to you at every turn; they are common in all the Islands; they are visible every day, and frequently at night also — not the silvery bow we see once in an age in the States, by moonlight, but barred with all bright and beautiful colors, like the children of the sun and rain. I saw one of them a few nights ago. What the sailors call "rain-dogs" — little patches of rainbow — are often seen drifting about the heavens in these latitudes, like stained cathedral windows.

Kealakekua Bay is a little curve like the last kink
of a snail-shell, winding deep into the land, seemingly not more than a mile wide from shore to shore.
It is bounded on one side — where the murder was
done — by a little flat plain, on which stands a
cocoanut grove and some ruined houses; a steep
wall of lava, a thousand feet high at the upper end
and three or four hundred at the lower, comes down
from the mountain and bounds the inner extremity
of it. From this wall the place takes its name,
Kealakekua, which in the native tongue signifies
"The Pathway of the Gods." They say (and still
believe, in spite of their liberal education in Christianity), that the great god *Lono*, who used to live
upon the hillside, always traveled that causeway
when urgent business connected with heavenly affairs
called him down to the seashore in a hurry.

As the red sun looked across the placid ocean
through the tall, clean stems of the cocoanut trees,
like a blooming whisky bloat through the bars of a
city prison, I went and stood in the edge of the
water on the flat rock pressed by Captain Cook's
feet when the blow was dealt which took away his
life, and tried to picture in my mind the doomed
man struggling in the midst of the multitude of
exasperated savages — the men in the ship crowding
to the vessel's side and gazing in anxious dismay
toward the shore — the — but I discovered that I
could not do it.

It was growing dark, the rain began to fall, we

R**

could see that the distant *Boomerang* was helplessly becalmed at sea, and so I adjourned to the cheerless little box of a warehouse and sat down to smoke and think, and wish the ship would make the land — for we had not eaten much for ten hours and were viciously hungry.

Plain unvarnished history takes the romance out of Captain Cook's assassination, and renders a deliberate verdict of justifiable homicide. Wherever he went among the islands, he was cordially received and welcomed by the inhabitants, and his ships lavishly supplied with all manner of food. He returned these kindnesses with insult and ill-treatment. Perceiving that the people took him for the long vanished and lamented god Lono, he encouraged them in the delusion for the sake of the limitless power it gave him; but during the famous disturbance at this spot, and while he and his comrades were surrounded by fifteen thousand maddened savages, he received a hurt and betrayed his earthly origin with a groan. It was his death-warrant. Instantly a shout went up: "He groans! — he is not a god!" So they closed in upon him and dispatched him.

His flesh was stripped from the bones and burned (except nine pounds of it which were sent on board the ships). The heart was hung up in a native hut, where it was found and eaten by three children, who mistook it for the heart of a dog. One of these children grew to be a very old man, and died in

Honolulu a few years ago. Some of Cook's bones were recovered and consigned to the deep by the officers of the ships.

Small blame should attach to the natives for the killing of Cook. They treated him well. In return, he abused them. He and his men inflicted bodily injury upon many of them at different times, and killed at least three of them before they offered any proportionate retaliation.

Near the shore we found "Cook's monument"— only a cocoanut stump, four feet high and about a foot in diameter at the butt. It had lava boulders piled around its base to hold it up and keep it in its place, and it was entirely sheathed over, from top to bottom, with rough, discolored sheets of copper, such as ships' bottoms are coppered with. Each sheet had a rude inscription scratched upon it— with a nail, apparently— and in every case the execution was wretched. Most of these merely recorded the visits of British naval commanders to the spot, but one of them bore this legend:

"Near this spot fell
CAPTAIN JAMES COOK,
The Distinguished Circumnavigator, who Discovered these Islands A. D. 1778."

After Cook's murder, his second in command, on board the ship, opened fire upon the swarms of natives on the beach, and one of his cannon-balls cut this cocoanut tree short off and left this monumental stump standing. It looked sad and lonely enough to us, out there in the rainy twilight. But there is no other monument to Captain Cook. True, up on the mountain side we had passed by a large inclosure like an ample hog-pen, built of lava blocks, which marks the spot where Cook's flesh was stripped from his bones and burned; but this is not properly a monument, since it was erected by the natives themselves, and less to do honor to the circumnavigator than for the sake of convenience in roasting him. A thing like a guideboard was elevated above this pen on a tall pole, and formerly there was an inscription upon it describing the memorable occurrence that had there taken place; but the sun and the wind have long ago so defaced it as to render it illegible.

Toward midnight a fine breeze sprang up and the schooner soon worked herself into the bay and cast anchor. The boat came ashore for us, and in a little while the clouds and the rain were all gone. The moon was beaming tranquilly down on land and sea, and we two were stretched upon the deck sleeping the refreshing sleep and dreaming the happy dreams that are only vouchsafed to the weary and the innocent.

CHAPTER XXXI.

IN the breezy morning we went ashore and visited the ruined temple of the last god Lono. The high chief cook of this temple — the priest who presided over it and roasted the human sacrifices — was uncle to Obookia, and at one time that youth was an apprentice-priest under him. Obookia was a young native of fine mind, who, together with three other native boys, was taken to New England by the captain of a whaleship during the reign of Kamehameha I., and they were the means of attracting the attention of the religious world to their country. This resulted in the sending of missionaries there. And this Obookia was the very same sensitive savage who sat down on the church steps and wept because his people did not have the Bible. That incident has been very elaborately painted in many a charming Sunday-school book — aye, and told so plaintively and so tenderly that I have cried over it in Sunday-school myself, on general principles, although at a time when I did not know much and could not understand why the people of the Sandwich Islands needed to worry so much about it as long as they did not know there was a Bible at all.

Obookia was converted and educated, and was to have returned to his native land with the first missionaries, had he lived. The other native youths made the voyage, and two of them did good service, but the third, William Kanui, fell from grace afterward, for a time, and when the gold excitement broke out in California he journeyed thither and went to mining, although he was fifty years old. He succeeded pretty well, but the failure of Page, Bacon & Co. relieved him of six thousand dollars, and then, to all intents and purposes, he was a bankrupt in his old age and he resumed service in the pulpit again. He died in Honolulu in 1864.

Quite a broad tract of land near the temple, extending from the sea to the mountain top, was sacred to the god Lono in olden times — so sacred that if a common native set his sacrilegious foot upon it, it was judicious for him to make his will, because his time had come. He might go around it by water, but he could not cross it. It was well sprinkled with pagan temples and stocked with awkward, homely idols carved out of logs of wood. There was a temple devoted to prayers for rain — and with fine sagacity it was placed at a point so well up on the mountain side that if you prayed there twenty-four times a day for rain you would be likely to get it every time. You would seldom get to your Amen before you would have to hoist your umbrella.

And there was a large temple near at hand which was built in a single night, in the midst of storm and thunder and rain, by the ghastly hands of dead men! Tradition says that by the weird glare of the lightning a noiseless multitude of phantoms were seen at their strange labor far up the mountain-side at dead of night — flitting hither and thither and bearing great lava-blocks clasped in their nerveless fingers — appearing and disappearing as the pallid luster fell upon their forms and faded away again. Even to this day, it is said, the natives hold this dread structure in awe and reverence, and will not pass by it in the night.

At noon I observed a bevy of nude native young ladies bathing in the sea, and went and sat down on their clothes to keep them from being stolen. I begged them to come out, for the sea was rising, and I was satisfied that they were running some risk. But they were not afraid, and presently went on with their sport. They were finished swimmers and divers, and enjoyed themselves to the last degree. They swam races, splashed and ducked and tumbled each other about, and filled the air with their laughter. It is said that the first thing an Islander learns is how to swim; learning to walk, being a matter of smaller consequence, comes afterward. One hears tales of native men and women swimming ashore from vessels many miles at sea — more miles, indeed, than I dare vouch for or even mention. And they tell of a native diver who went down in thirty or forty-foot waters and brought up an anvil! I think he swallowed the anvil afterward, if my memory serves me. However, I will not urge this point.

I have spoken several times of the god Lono—I may as well furnish two or three sentences concerning him.

The idol the natives worshiped for him was a slender, unornamented staff twelve feet long. Tradition says he was a favorite god on the Island of Hawaii—a great king who had been deified for meritorious services—just our own fashion of rewarding heroes, with the difference that we would have made him a postmaster instead of a god, no doubt. In an angry moment he slew his wife, a goddess named Kaikilani Aiii. Remorse of conscience drove him mad, and tradition presents us the singular spectacle of a god traveling "on the shoulder"; for in his gnawing grief he wandered about from place to place boxing and wrestling with all whom he met. Of course, this pastime soon lost its novelty, inasmuch as it must necessarily have been the case that when so powerful a deity sent a frail human opponent "to grass" he never came back any more. Therefore, he instituted games called makahiki, and ordered that they should be held in his honor, and then sailed for foreign lands on a three-cornered raft, stating that he would return some day—and that was the last of Lono. He was never seen any more; his raft got swamped, perhaps. But the people always expected his return, and thus they were easily led to accept Captain Cook as the restored god.

Some of the old natives believed Cook was Lono to the day of their death; but many did not, for they could not understand how he could die if he was a god.

Only a mile or so from Kealakekua Bay is a spot of historic interest — the place where the last battle was fought for idolatry. Of course, we visited it, and came away as wise as most people do who go and gaze upon such mementoes of the past when in an unreflective mood.

While the first missionaries were on their way around the Horn, the idolatrous customs which had obtained in the island, as far back as tradition reached, were suddenly broken up. Old Kamehameha I. was dead, and his son, Liholiho, the new king, was a free liver, a roystering, dissolute fellow, and hated the restraints of the ancient *tabu*. His assistant in the government, Kaahumanu, the queen dowager, was proud and high-spirited, and hated the *tabu* because it restricted the privileges of her sex and degraded all women very nearly to the level of brutes. So the case stood. Liholiho had half a mind to put his foot down, and Kaahumanu had a whole mind to badger him into doing it, and whisky did the rest. It was probably the first time whisky ever prominently figured as an aid to civilization. Liholiho came up to Kailua as drunk as a piper, and attended a great feast; the determined queen spurred his drunken courage up to a reckless pitch, and then, while all the multitude

stared in blank dismay, he moved deliberately forward and sat down with the women! They saw
him eat from the same vessel with them, and were
appalled! Terrible moments drifted slowly by,
still the king ate, still he lived, still the lightnings of
the insulted gods were withheld! Then conviction
came like a revelation — the superstitions of a hundred generations passed from before the people like
a cloud, and a shout went up, "The *tabu* is broken!
The *tabu* is broken!"

Thus did King Liholiho and his dreadful whisky
preach the first sermon and prepare the way for the
new gospel that was speeding southward over the
waves of the Atlantic.

The *tabu* broken and destruction failing to follow
the awful sacrilege, the people, with that childlike
precipitancy which has always characterized them,
jumped to the conclusion that their gods were a
weak and wretched swindle, just as they formerly
jumped to the conclusion that Captain Cook was no
god, merely because he groaned, and promptly
killed him without stopping to inquire whether a
god might not groan as well as a man if it suited his
convenience to do it; and satisfied that the idols
were powerless to protect themselves they went to
work at once and pulled them down — hacked them
to pieces — applied the torch — annihilated them!

The pagan priests were furious. And well they
might be; they had held the fattest offices in the
land, and now they were beggared; they had been

great — they had stood above the chiefs — and now they were vagabonds. They raised a revolt; they scared a number of people into joining their standard, and Bekuokalani, an ambitious offshoot of royalty, was easily persuaded to become their leader.

In the first skirmish the idolaters triumphed over the royal army sent against them, and full of confidence they resolved to march upon Kailua. The king sent an envoy to try and conciliate them, and came very near being an envoy short by the operation; the savages not only refused to listen to him, but wanted to kill them. So the king sent his men forth under Major-General Kalaimoku and the two hosts met at Kuamoo. The battle was long and fierce — men and women fighting side by side, as was the custom — and when the day was done the rebels were flying in every direction in hopeless panic, and idolatry and the *tabu* were dead in the land!

The royalists marched gayly home to Kailua glorifying the new dispensation. "There is no power in the gods," said they; "they are a vanity and a lie. The army with idols was weak; the army without idols was strong and victorious!"

The nation was without a religion.

The missionary ship arrived in safety shortly afterward, timed by providential exactness to meet the emergency, and the gospel was planted as in a virgin soil.

CHAPTER XXXII.

AT noon, we hired a Kanaka to take us down to the ancient ruins at Honaunau in his canoe — price two dollars — reasonable enough, for a sea voyage of eight miles, counting both ways.

The native canoe is an irresponsible looking contrivance. I cannot think of anything to liken it to but a boy's sled-runner hollowed out, and that does not quite convey the correct idea. It is about fifteen feet long, high and pointed at both ends, is a foot and a half or two feet deep, and so narrow that if you wedged a fat man into it you might not get him out again. It sits on top of the water like a duck, but it has an outrigger and does not upset easily, if you keep still. This outrigger is formed of two long bent sticks like plow-handles, which project from one side, and to their outer ends is bound a curved beam composed of an extremely light wood, which skims along the surface of the water, and thus saves you from an upset on that side, while the outrigger's weight is not so easily lifted as to make an upset on the other side a thing to be greatly feared. Still, until one gets used to sitting perched upon this knifeblade, he is apt to reason within himself that it

would be more comfortable if there were just an
outrigger or so on the other side also.

I had the bow seat, and Billings sat amidships and
faced the Kanaka, who occupied the stern of the
craft and did the paddling. With the first stroke
the trim shell of a thing shot out from the shore
like an arrow. There was not much to see. While
we were on the shallow water of the reef, it was
pastime to look down into the limpid depths at the
large bunches of branching coral — the unique
shrubbery of the sea. We lost that, though, when
we got out into the dead blue water of the deep.
But we had the picture of the surf, then, dashing
angrily against the crag-bound shore and sending a
foaming spray high into the air. There was interest
in this beetling border, too, for it was honeycombed
with quaint caves and arches and tunnels, and had a
rude semblance of the dilapidated architecture of
ruined keeps and castles rising out of the restless
sea. When this novelty ceased to be a novelty, we
turned our eyes shoreward and gazed at the long
mountain with its rich green forests stretching up
into the curtaining clouds, and at the specks of
houses in the rearward distance and the diminished
schooner riding sleepily at anchor. And when these
grew tiresome we dashed boldly into the midst of a
school of huge, beastly porpoises engaged at their
eternal game of arching over a wave and disappearing, and then doing it over again and keeping it up
— always circling over, in that way, like so many

well-submerged wheels. But the porpoises wheeled themselves away, and then we were thrown upon our own resources. It did not take many minutes to discover that the sun was blazing like a bonfire, and that the weather was of a melting temperature. It had a drowsing effect, too.

In one place we came upon a large company of naked natives, of both sexes and all ages, amusing themselves with the national pastime of surf-bathing. Each heathen would paddle three or four hundred yards out to sea (taking a short board with him), then face the shore and wait for a particularly pro-digious billow to come along; at the right moment he would fling his board upon its foamy crest and himself upon the board, and here he would come whizzing by like a bombshell! It did not seem that a lightning express train could shoot along at a more hair-lifting speed. I tried surf-bathing once, subse-quently, but made a failure of it. I got the board placed right, and at the right moment, too; but missed the connection myself. The board struck the shore in three-quarters of a second, without any cargo, and I struck the bottom about the same time, with a couple of barrels of water in me. None but natives ever master the art of surf-bathing thor-oughly.

At the end of an hour, we had made the four miles, and landed on a level point of land, upon which was a wide extent of old ruins, with many a tall cocoanut tree growing among them. Here was the ancient City of Refuge — a vast inclosure, whose stone walls were twenty feet thick at the base, and fifteen feet high; an oblong square, a thousand and forty feet one way and a fraction under seven hundred the other. Within this inclosure, in early times, have been three rude temples, each two hundred and ten feet long by one hundred wide, and thirteen high.

In those days, if a man killed another anywhere on the Island the relatives were privileged to take the murderer's life; and then a chase for life and liberty began — the outlawed criminal flying through pathless forests and over mountain and plain, with his hopes fixed upon the protecting walls of the City of Refuge, and the avenger of blood following hotly after him! Sometimes the race was kept up to the very gates of the temple, and the panting pair sped through long files of excited natives, who watched the contest with flashing eye and dilated nostril, encouraging the hunted refugee with sharp, inspiriting ejaculations, and sending up a ringing shout of exultation when the saving gates closed upon him and the cheated pursuer sank exhausted at the threshold. But sometimes the flying criminal fell under the hand of the avenger at the very door, when one more brave stride, one more brief second of time would have brought his feet upon the sacred ground and barred him against all harm. Where did these isolated pagans get this idea of a City of Refuge — this ancient Oriental custom?

19**

This old sanctuary was sacred to all — even to
rebels in arms and invading armies. Once within
its walls, and confession made to the priest and
absolution obtained, the wretch with a price upon
his head could go forth without fear and without
danger — he was *tabu*, and to harm him was death.
The routed rebels in the lost battle for idolatry fled
to this place to claim sanctuary, and many were
thus saved.

Close to the corner of the great inclosure is a
round structure of stone, some six or eight feet
high, with a level top about ten or twelve in
diameter. This was the place of execution. A
high palisade of cocoanut piles shut out the cruel
scenes from the vulgar multitude. Here criminals
were killed, the flesh stripped from the bones and
burned, and the bones secreted in holes in the body
of the structure. If the man had been guilty of a
high crime, the entire corpse was burned.

The walls of the temple are a study. The same
food for speculation that is offered the visitor to the
Pyramids of Egypt he will find here — the mystery
of how they were constructed by a people unacquainted with science and mechanics. The natives
have no invention of their own for hoisting heavy
weights, they had no beasts of burden, and they
have never even shown any knowledge of the prop-
erties of the lever. Yet some of the lava blocks
quarried out, brought over rough, broken ground,
and built into this wall, six or seven feet from the

ground, are of prodigious size and would weigh
tons. How did they transport and how raise them?

Both the inner and outer surfaces of the walls
present a smooth front and are very creditable
specimens of masonry. The blocks are of all manner of shapes and sizes, but yet are fitted together
with the neatest exactness. The gradual narrowing
of the wall from the base upward is accurately
preserved.

No cement was used, but the edifice is firm and
compact and is capable of resisting storm and decay
for centuries. Who built this temple, and how was
it built, and when, are mysteries that may never be
unraveled.

Outside of these ancient walls lies a sort of coffin-
shaped stone eleven feet four inches long and three
feet square at the small end (it would weigh a few
thousand pounds), which the high chief who held
sway over this district many centuries ago brought
thither on his shoulder one day to use as a lounge!
This circumstance is established by the most reliable
traditions. He used to lie down on it, in his indolent way, and keep an eye on his subjects at work
for him and see that there was no "soldiering"
done. And no doubt there was not any done to
speak of, because he was a man of that sort of build
that incites to attention to business on the part of
an employé. He was fourteen or fifteen feet high.
When he stretched himself at full length on his
lounge, his legs hung down over the end, and when

S**

he snored he woke the dead. These facts are all attested by irrefragable tradition.

On the other side of the temple is a monstrous seven-ton rock, eleven feet long, seven feet wide, and three feet thick. It is raised a foot or a foot and a half above the ground, and rests upon half-a-dozen little stony pedestals. The same old fourteen-footer brought it down from the mountain, merely for fun (he had his own notions about fun), and propped it up as we find it now and as others may find it a century hence, for it would take a score of horses to budge it from its position. They say that fifty or sixty years ago the proud Queen Kaahumanu used to fly to this rock for safety, whenever she had been making trouble with her fierce husband, and hide under it until his wrath was appeased. But these Kanakas will lie, and this statement is one of their ablest efforts — for Kaahumanu was six feet high — she was bulky — she was built like an ox — and she could no more have squeezed herself under that rock than she could have passed between the cylinders of a sugar mill. What could she gain by it, even if she succeeded? To be chased and abused by a savage husband could not be otherwise than humiliating to her high spirit, yet it could never make her feel so flat as an hour's repose under that rock would.

We walked a mile over a raised macadamized road of uniform width; a road paved with flat stones and exhibiting in its every detail a considerable degree of engineering skill. Some say that that wise old pagan, Kamehameha I, planned and built it, but others say it was built so long before his time that the knowledge of who constructed it has passed out of the traditions. In either case, however, as the handiwork of an untaught and degraded race it is a thing of pleasing interest. The stones are worn and smooth, and pushed apart in places, so that the road has the exact appearance of those ancient paved highways leading out of Rome which one sees in pictures.

The object of our tramp was to visit a great natural curiosity at the base of the foothills — a congealed cascade of lava. Some old forgotten volcanic eruption sent its broad river of fire down the mountain-side here, and it poured down in a great torrent from an overhanging bluff some fifty feet high to the ground below. The flaming torrent cooled in the winds from the sea, and remains there to-day, all seamed, and frothed, and rippled, a petrified Niagara. It is very picturesque, and withal so natural that one might almost imagine it still flowed. A smaller stream trickled over the cliff and built up an isolated pyramid about thirty feet high, which has the semblance of a mass of large gnarled and knotted vines and roots and stems intricately twisted and woven together.

We passed in behind the cascade and the pyramid, and found the bluff pierced by several cavernous tunnels, whose crooked courses we followed a long distance.

Two of these winding tunnels stand as proof of nature's mining abilities. Their floors are level, they are seven feet wide, and their roofs are gently arched. Their height is not uniform, however. We passed through one a hundred feet long, which leads through a spur of the hill and opens out well up in the sheer wall of a precipice whose foot rests in the waves of the sea. It is a commodious tunnel, except that there are occasional places in it where one must stoop to pass under. The roof is lava, of course, and is thickly studded with little lava-pointed icicles an inch long, which hardened as they dripped. They project as closely together as the iron teeth of a corn-sheller, and if one will stand up straight and walk any distance there, he can get his hair combed free of charge.

CHAPTER XXXIII.

WE got back to the schooner in good time, and then sailed down to Kau, where we disembarked and took final leave of the vessel. Next day we bought horses and bent our way over the summer-clad mountain-terraces, toward the great volcano of Kilauea (Ke-low-way-ah). We made nearly a two-days' journey of it, but that was on account of laziness. Toward sunset on the second day, we reached an elevation of some four thousand feet above sea level, and as we picked our careful way through billowy wastes of lava long generations ago stricken dead and cold in the climax of its tossing fury, we began to come upon signs of the near presence of the volcano — signs in the nature of ragged fissures that discharged jets of sulphurous vapor into the air, hot from the molten ocean down in the bowels of the mountain.

Shortly the crater came into view. I have seen Vesuvius since, but it was a mere toy, a child's volcano, a soup-kettle, compared to this. Mount Vesuvius is a shapely cone thirty-six hundred feet high; its crater an inverted cone only three hundred feet deep, and not more than a thousand feet in diameter, if as much as that; its fires meager, modest, and docile. But here was a vast, perpendicular, walled cellar, nine hundred feet deep in some places, thirteen hundred in others, level-floored, and *ten miles in circumference!* Here was a yawning pit upon whose floor the armies of Russia could camp, and have room to spare.

Perched upon the edge of the crater, at the opposite end from where we stood, was a small lookout house — say three miles away. It assisted us, by comparison, to comprehend and appreciate the great depth of the basin — it looked like a tiny martin-box clinging at the eaves of a cathedral. After some little time spent in resting and looking and ciphering, we hurried on to the hotel.

By the path it is half a mile from the Volcano House to the lookout-house. After a hearty supper we waited until it was thoroughly dark and then started to the crater. The first glance in that direction revealed a scene of wild beauty. There was a heavy fog over the crater and it was splendidly illuminated by the glare from the fires below. The illumination was two miles wide and a mile high, perhaps; and if you ever, on a dark night and at a distance, beheld the light from thirty or forty blocks of distant buildings all on fire at once, reflected strongly against overhanging clouds, you can form a fair idea of what this looked like.

A colossal column of cloud towered to a great

height in the air immediately above the crater, and
the outer swell of every one of its vast folds was
dyed with a rich crimson luster, which was subdued
to a pale rose tint in the depressions between. It
glowed like a muffled torch and stretched upward to
a dizzy height toward the zenith. I thought it just
possible that its like had not been seen since the
children of Israel wandered on their long march
through the desert so many centuries ago over a
path illuminated by the mysterious " pillar of fire."
And I was sure that I now had a vivid conception
of what the majestic " pillar of fire " was like, which
almost amounted to a revelation.

Arrived at the little thatched lookout-house, we
rested our elbows on the railing in front and looked
abroad over the wide crater and down over the sheer
precipice at the seething fires beneath us. The view
was a startling improvement on my daylight experience. I turned to see the effect on the balance of
the company, and found the reddest-faced set of men
I almost ever saw. In the strong light every countenance glowed like red-hot iron, every shoulder
was suffused with crimson and shaded rearward into
dingy, shapeless obscurity! The place below looked
like the infernal regions and these men like half-
cooled devils just come up on a furlough.

I turned my eyes upon the volcano again. The
" cellar " was tolerably well lighted up. For a mile
and a half in front of us and half a mile on either
side, the floor of the abyss was magnificently illu-

minated; beyond these limits the mists hung down their gauzy curtains and cast a deceptive gloom over all that made the twinkling fires in the remote corners of the crater seem countless leagues removed — made them seem like the camp-fires of a great army far away. Here was room for the imagination to work! You could imagine those lights the width of a continent away — and that hidden under the intervening darkness were hills, and winding rivers, and weary wastes of plain and desert — and even then the tremendous vista stretched on, and on, and on! — to the fires and far beyond! You could not compass it — it was the idea of eternity made tangible — and the longest end of it made visible to the naked eye!

The greater part of the vast floor of the desert under us was as black as ink, and apparently smooth and level; but over a mile square of it was ringed and streaked and striped with a thousand branching streams of liquid and gorgeously brilliant fire! It looked like a colossal railroad map of the State of Massachusetts done in chain lightning on a midnight sky. Imagine it — imagine a coal-black sky shivered into a tangled network of angry fire!

Here and there were gleaming holes a hundred feet in diameter, broken in the dark crust, and in them the melted lava — the color a dazzling white just tinged with yellow — was boiling and surging furiously; and from these holes branched numberless bright torrents in many directions, like the spokes of a wheel, and kept a tolerably straight course for a while and then swept round in huge rainbow curves, or made a long succession of sharp worm-fence angles, which looked precisely like the fiercest jagged lightning. These streams met other streams, and they mingled with and crossed and recrossed each other in every conceivable direction, like skate tracks on a popular skating ground. Sometimes streams twenty or thirty feet wide flowed from the holes to some distance without dividing — and through the opera-glasses we could see that they ran down small, steep hills and were genuine cataracts of fire, white at their source, but soon cooling and turning to the richest red, grained with alternate lines of black and gold. Every now and then masses of the dark crust broke away and floated slowly down these streams like rafts down a river. Occasionally, the molten lava flowing under the superincumbent crust broke through — split a dazzling streak, from five hundred to a thousand feet long, like a sudden flash of lightning, and then acre after acre of the cold lava parted into fragments, turned up edgewise like cakes of ice when a great river breaks up, plunged downward and were swallowed in the crimson cauldron. Then the wide expanse of the "thaw" maintained a ruddy glow for a while, but shortly cooled and became black and level again. During a "thaw," every dismembered cake was marked by a glittering white border which was superbly shaded inward by aurora borealis rays,

which were a flaming yellow where they joined the
white border, and from thence toward their points
tapered into glowing crimson, then into a rich, pale
carmine, and finally into a faint blush that held its
own a moment and then dimmed and turned black.
Some of the streams preferred to mingle together in
a tangle of fantastic circles, and then they looked
something like the confusion of ropes one sees on a
ship's deck when she has just taken in sail and
dropped anchor — provided one can imagine those
ropes on fire.

Through the glasses, the little fountains scattered
about looked very beautiful. They boiled, and
coughed, and spluttered, and discharged sprays of
stringy red fire — of about the consistency of mush,
for instance — from ten to fifteen feet into the air,
along with a shower of brilliant white sparks — a
quaint and unnatural mingling of gouts of blood and
snowflakes!

We had circles and serpents and streaks of lightning all twined and wreathed and tied together,
without a break throughout an area more than a mile
square (that amount of ground was covered, though
it was not strictly " square "), and it was with a
feeling of placid exultation that we reflected that
many years had elapsed since any visitor had seen
such a splendid display — since any visitor had seen
anything more than the now snubbed and insignificant " North " and " South " lakes in action. We
had been reading old files of Hawaiian newspapers

and the "Record Book" at the Volcano House, and were posted.

I could see the North Lake lying out on the black floor away off in the outer edge of our panorama, and knitted to it by a web-work of lava streams. In its individual capacity it looked very little more respectable than a schoolhouse on fire. True, it was about nine hundred feet long and two or three hundred wide, but then, under the present circumstances, it necessarily appeared rather insignificant, and besides it was so distant from us.

I forgot to say that the noise made by the bubbling lava is not great, heard as we heard it from our lofty perch. It makes three distinct sounds — a rushing, a hissing, and a coughing or puffing sound, and if you stand on the brink and close your eyes, it is no trick at all to imagine that you are sweeping down a river on a large low-pressure steamer, and that you hear the hissing of the steam about her boilers, the puffing from her escape-pipes and the churning rush of the water abaft her wheels. The smell of sulphur is strong, but not unpleasant to a sinner.

We left the lookout-house at ten o'clock in a half cooked condition, because of the heat from Pele's furnaces, and, wrapping up in blankets, for the night was cold, we returned to our hotel.

CHAPTER XXXIV.

THE next night was appointed for a visit to the bottom of the crater, for we desired to traverse its floor and see the " North Lake " (of fire) which lay two miles away, toward the further wall. After dark half a dozen of us set out, with lanterns and native guides, and climbed down a crazy, thousand-foot pathway in a crevice fractured in the crater wall, and reached the bottom in safety.

The irruption of the previous evening had spent its force and the floor looked black and cold; but when we ran out upon it we found it hot yet, to the feet, and it was likewise riven with crevices which revealed the underlying fires gleaming vindictively. A neighboring cauldron was threatening to overflow, and this added to the dubiousness of the situation. So the native guides refused to continue the venture, and then everybody deserted except a stranger named Marlette. He said he had been in the crater a dozen times in daylight and believed he could find his way through it at night. He thought that a run of three hundred yards would carry us over the hottest part of the floor and leave us our shoe-soles.

His pluck gave me backbone.   We took one lantern
and instructed the guides to hang the other to the
roof of the lookout-house to serve as a beacon for
us in case we got lost, and then the party started
back up the precipice and Marlette and I made our
run.   We skipped over the hot floor and over the
red crevices with brisk dispatch and reached the cold
lava safe but with pretty warm feet.   Then we took
things leisurely and comfortably, jumping tolerably
wide and probably bottomless chasms, and thread-
ing our way through picturesque lava upheavals with
considerable confidence.   When we got fairly away
from the cauldrons of boiling fire, we seemed to be
in a gloomy desert, and a suffocatingly dark one,
surrounded by dim walls that seemed to tower to the
sky.   The only cheerful objects were the glinting
stars high overhead.

By and by Marlette shouted "Stop!"   I never
stopped quicker in my life.   I asked what the
matter was.   He said we were out of the path.   He
said we must not try to go on till we found it again,
for we were surrounded with beds of rotten lava
through which we could easily break and plunge
down a thousand feet.   I thought eight hundred
would answer for me, and was about to say so when
Marlette partly proved his statement by accidentally
crushing through and disappearing to his armpits.
He got out and we hunted for the path with the
lantern.   He said there was only one path, and that
it was but vaguely defined.   We could not find it.

20 **

The lava surface was all alike in the lantern light.
But he was an ingenious man. He said it was not
the lantern that had informed him that we were out
of the path, but his *feet*. He had noticed a crisp
grinding of fine lava-needles under his feet, and
some instinct reminded him that in the path these
were all worn away. So he put the lantern behind
him, and began to search with his boots instead of
his eyes. It was good sagacity. The first time his
foot touched a surface that did not grind under it he
announced that the trail was found again; and after
that we kept up a sharp listening for the rasping
sound, and it always warned us in time.

It was a long tramp, but an exciting one. We
reached the North Lake between ten and eleven
o'clock, and sat down on a huge overhanging lava-
shelf, tired but satisfied. The spectacle presented
was worth coming double the distance to see.
Under us, and stretching away before us, was a
heaving sea of molten fire of seemingly limitless extent. The glare from it was so blinding that it was
some time before we could bear to look upon it
steadily. It was like gazing at the sun at noonday,
except that the glare was not quite so white. At
unequal distances all around the shores of the lake
were nearly white-hot chimneys or hollow drums of
lava, four or five feet high, and up through them
were bursting gorgeous sprays of lava-gouts and gem
spangles, some white, some red, and some golden —
a ceaseless bombardment, and one that fascinated

the eye with its unapproachable splendor. The
more distant jets, sparkling up through an intervening gossamer veil of vapor, seemed miles away; and
the further the curving ranks of fiery fountains re-
ceded, the more fairy-like and beautiful they appeared.

Now and then the surging bosom of the lake
under our noses would calm down ominously and
seem to be gathering strength for an enterprise; and
then all of a sudden a red dome of lava of the bulk
of an ordinary dwelling would heave itself aloft like
an escaping balloon, then burst asunder, and out of
its heart would flit a pale-green film of vapor, and
float upward and vanish in the darkness — a released
soul soaring homeward from captivity with the
damned, no doubt. The crashing plunge of the
ruined dome into the lake again would send a world
of seething billows lashing against the shores and
shaking the foundations of our perch. By and by,
a loosened mass of the hanging shelf we sat on
tumbled into the lake, jarring the surroundings like
an earthquake and delivering a suggestion that may
have been intended for a hint, and may not. We
did not wait to see.

We got lost again on our way back, and were
more than an hour hunting for the path. We were
where we could see the beacon lantern at the lookout-
house at the time, but thought it was a star, and paid
no attention to it. We reached the hotel at two
o'clock in the morning, pretty well fagged out.

20 **

Kilauea never overflows its vast crater, but bursts
a passage for its lava through the mountain-side
when relief is necessary, and then the destruction is
fearful.   About 1840 it rent its overburdened
stomach and sent a broad river of fire careering
down to the sea, which swept away forests, huts,
plantations, and everything else that lay in its path.
The stream was *five miles broad*, in places, and *two
hundred feet deep*, and the distance it traveled was
forty miles.   It tore up and bore away acre-patches
of land on its bosom like rafts — rocks, trees, and
all intact.   At night the red glare was visible a hundred miles at sea; and at a distance of forty miles
fine print could be read at midnight.   The atmosphere was poisoned with sulphurous vapors and
choked with falling ashes, pumice stones, and
cinders; countless columns of smoke rose up and
blended together in a tumbled canopy that hid the
heavens and glowed with a ruddy flush reflected
from the fires below; here and there jets of lava
sprung hundreds of feet into the air and burst into
rocket-sprays that returned to earth in a crimson
rain; and all the while the laboring mountain shook
with nature's great palsy, and voiced its distress in
moanings and the muffled booming of subterranean
thunders.

Fishes were killed for twenty miles along the
shore, where the lava entered the sea.   The earth-
quakes caused some loss of human life, and a prodigious tidal-wave swept inland, carrying everything

before it and drowning a number of natives. The devastation consummated along the route traversed by the river of lava was complete and incalculable. Only a Pompeii and a Herculaneum were needed at the foot of Kilauea to make the story of the irruption immortal.

CHAPTER XXXV.

WE rode horseback all around the island of
Hawaii (the crooked road making the distance two hundred miles), and enjoyed the journey
very much. We were more than a week making the
trip, because our Kanaka horses would not go by a
house or a hut without stopping — whip and spur
could not alter their minds about it, and so we
finally found that it economized time to let them
have their way. Upon inquiry the mystery was
explained; the natives are such thorough-going
gossips that they never pass a house without stopping to swap news, and consequently their horses
learn to regard that sort of thing as an essential part
of the whole duty of man, and his salvation not to
be compassed without it. However, at a former
crisis of my life I had once taken an aristocratic
young lady out driving, behind a horse that had just
retired from a long and honorable career as the
moving impulse of a milk wagon, and so this present
experience awoke a reminiscent sadness in me in
place of the exasperation more natural to the occasion. I remembered how helpless I was that day,

and how humiliated; how ashamed I was of having
intimated to the girl that I had always owned the
horse and was accustomed to grandeur; how hard I
tried to appear easy, and even vivacious, under
suffering that was consuming my vitals; how placidly
and maliciously the girl smiled, and kept on smiling,
while my hot blushes baked themselves into a permanent blood-pudding in my face; how the horse
ambled from one side of the street to the other and
waited complacently before every third house two
minutes and a quarter while I belabored his back
and reviled him in my heart; how I tried to keep
him from turning corners, and failed; how I moved
heaven and earth to get him out of town, and did
not succeed; how he traversed the entire settlement
and delivered imaginary milk at a hundred and sixty-
two different domiciles, and how he finally brought
up at a dairy depot and refused to budge further,
thus rounding and completing the revealment of
what the plebeian service of his life had been; how,
in eloquent silence, I walked the girl home, and how,
when I took leave of her, her parting remark
scorched my soul and appeared to blister me all
over; she said that my horse was a fine, capable
animal, and I must have taken great comfort in him
in my time — but that if I would take along some
milk-tickets next time, and appear to deliver them
at the various halting places, it might expedite his
movements a little. There was a coolness between
us after that.

In one place in the island of Hawaii, we saw a laced and ruffled cataract of limpid water leaping from a sheer precipice fifteen hundred feet high; but that sort of scenery finds its stanchest ally in the arithmetic rather than in spectacular effect. If one desires to be so stirred by a poem of nature wrought in the happily commingled graces of picturesque rocks, glimpsed distances, foliage, color, shifting lights and shadows, and falling water, that the tears almost come into his eyes so potent is the charm exerted, he need not go away from America to enjoy such an experience. The Rainbow Fall, in Watkins Glen (N. Y.), on the Erie railway, is an example. It would recede into pitiable insignificance if the callous tourist drew an arithmetic on it; but left to compete for the honors simply on scenic grace and beauty — the grand, the august, and the sublime being barred the contest — it could challenge the old world and the new to produce its peer.

In one locality, on our journey, we saw some horses that had been born and reared on top of the mountains, above the range of running water, and consequently they had never drunk that fluid in their lives, but had been always accustomed to quenching their thirst by eating dew-laden or shower-wetted leaves. And now it was destructively funny to see them sniff suspiciously at a pail of water, and then put in their noses and try to take a *bite* out of the fluid, as if it were a solid. Finding it liquid, they would snatch away their heads and fall to trembling,

snorting, and showing other evidences of fright.
When they became convinced at last that the water
was friendly and harmless, they thrust in their noses
up to their eyes, brought out a mouthful of the
water, and proceeded to *chew* it complacently. We
saw a man coax, kick, and spur one of them five or
ten minutes before he could make it cross a running
stream. It spread its nostrils, distended its eyes,
and trembled all over, just as horses customarily do
in the presence of a serpent — and for aught I know
it thought the crawling stream *was* a serpent.

In due course of time our journey came to an end
at Kawaehae (usually pronounced To-a-*hi* — and
before we find fault with this elaborate orthograph-
ical method of arriving at such an unostentatious result, let us lop off the *ugh* from our word
"though"). I made this horseback trip on a mule.
I paid ten dollars for him at Kau (Kah-oo), added
four to get him shod, rode him two hundred miles,
and then sold him for fifteen dollars. I mark the
circumstance with a white stone (in the absence of
chalk — for I never saw a white stone that a body
could mark anything with, though out of respect
for the ancients I have tried it often enough) ; for
up to that day and date it was the first strictly com-
mercial transaction I had ever entered into, and
come out winner. We returned to Honolulu, and
from thence sailed to the island of Maui, and spent
several weeks there very pleasantly. I still remember, with a sense of indolent luxury, a picnicking

excursion up a romantic gorge there, called the Iao
Valley. The trail lay along the edge of a brawling
stream in the bottom of the gorge — a shady route,
for it was well roofed with the verdant domes of
forest trees. Through openings in the foliage we
glimpsed picturesque scenery that revealed ceaseless
changes and new charms with every step of our
progress. Perpendicular walls from one to three
thousand feet high guarded the way, and were
sumptuously plumed with varied foliage in places,
and in places swathed in waving ferns. Passing
shreds of cloud trailed their shadows across these
shining fronts, mottling them with blots; billowy
masses of white vapor hid the turreted summits, and
far above the vapor swelled a background of gleaming green crags and cones that came and went,
through the veiling mists, like islands drifting in a
fog; sometimes the cloudy curtain descended till
half the canyon wall was hidden, then shredded gradually away till only airy glimpses of the ferny front
appeared through it — then swept aloft and left it
glorified in the sun again. Now and then, as our
position changed, rocky bastions swung out from the
wall, a mimic ruin of castellated ramparts and crum-
bling towers clothed with mosses and hung with
garlands of swaying vines, and as we moved on they
swung back again and hid themselves once more in
the foliage. Presently, a verdure-clad needle of
stone, a thousand feet high, stepped out from behind a corner, and mounted guard over the mys-

teries of the valley. It seemed to me that if Captain Cook needed a monument, here was one ready
made — therefore, why not put up his sign here,
and sell out the venerable cocoanut stump?

But the chief pride of Maui is her dead volcano of
Haleakala — which means, translated, "the house
of the sun." We climbed a thousand feet up the
side of this isolated colossus one afternoon; then
camped, and next day climbed the remaining nine
thousand feet, and anchored on the summit, where
we built a fire and froze and roasted by turns, all
night. With the first pallor of dawn we got up and
saw things that were new to us. Mounted on a
commanding pinnacle, we watched nature work her
silent wonders. The sea was spread abroad on every
hand, its tumbled surface seeming only wrinkled and
dimpled in the distance. A broad valley below appeared like an ample checker-board, its velvety
green sugar plantations alternating with dun squares
of barrenness and groves of trees diminished to
mossy tufts. Beyond the valley were mountains
picturesquely grouped together; but, bear in mind,
we fancied that we were looking *up* at these things
— not down. We seemed to sit in the bottom of a
symmetrical bowl ten thousand feet deep, with the
valley and the skirting sea lifted away into the sky
above us! It was curious; and not only curious,
but aggravating; for it was having our trouble all
for nothing, to climb ten thousand feet toward
heaven and then have to look *up* at our scenery.

However, we had to be content with it and make the best of it; for, all we could do we could not coax our landscape down out of the clouds. Formerly, when I had read an article in which Poe treated of this singular fraud perpetrated upon the eye by isolated great altitudes, I had looked upon the matter as an invention of his own fancy.

I have spoken of the outside view — but we had an inside one, too. That was the yawning dead crater, into which we now and then tumbled rocks, half as large as a barrel, from our perch, and saw them go careering down the almost perpendicular sides, bounding three hundred feet at a jump; kicking up dust-clouds wherever they struck; diminishing to our view as they sped farther into distance; growing invisible, finally, and only betraying their course by faint little puffs of dust; and coming to a halt at last in the bottom of the abyss, two thousand five hundred feet down from where they started! It was magnificent sport. We wore ourselves out at it.

The crater of Vesuvius, as I have before remarked, is a modest pit about a thousand feet deep and three thousand in circumference; that of Kilauea is somewhat deeper, and *ten miles* in circumference. But what are either of them compared to the vacant stomach of Haleakala? I will not offer any figures of my own, but give official ones — those of Commander Wilkes, U. S. N., who surveyed it and testifies that it is *twenty-seven miles in circumference!* If it had a level bottom it would make a fine site for a city like London. It must have afforded a spectacle worth contemplating in the old days when its furnaces gave full rein to their anger.

Presently, vagrant white clouds came drifting along, high over the sea and the valley; then they came in couples and groups; then in imposing squadrons; gradually joining their forces, they banked themselves solidly together, a thousand feet under us, and *totally shut out land and ocean* — not a vestige of *anything* was left in view, but just a little of the rim of the crater, circling away from the pinnacle whereon we sat (for a ghostly procession of wanderers from the filmy hosts without had drifted through a chasm in the crater wall and filed round and round, and gathered and sunk and blended together till the abyss was stored to the brim with a fleecy fog). Thus banked, motion ceased, and silence reigned. Clear to the horizon, league on league, the snowy floor stretched without a break — not level, but in rounded folds, with shallow creases between, and with here and there stately piles of vapory architecture lifting themselves aloft out of the common plain — some near at hand, some in the middle distances, and others relieving the monotony of the remote solitudes. There was little conversation, for the impressive scene overawed speech. I felt like the Last Man, neglected of the judgment, and left pinnacled in mid-heaven, a forgotten relic of a vanished world.

While the hush yet brooded, the messengers of

the coming resurrection appeared in the East. A
growing warmth suffused the horizon, and soon the
sun emerged and looked out over the cloud-waste,
flinging bars of ruddy light across it, staining its
folds and billow - caps with blushes, purpling the
shaded troughs between, and glorifying the massy
vapor-palaces and cathedrals with a wasteful splendor
of all blendings and combinations of rich coloring.

It was the sublimest spectacle I ever witnessed,
and I think the memory of it will remain with me
always.

CHAPTER XXXVI.

I STUMBLED upon one curious character in the Island of Maui. He became a sore annoyance to me in the course of time. My first glimpse of him was in a sort of public room in the town of Lahaina. He occupied a chair at the opposite side of the apartment, and sat eying our party with interest for some minutes, and listening as critically to what we were saying as if he fancied we were talking to him and expecting him to reply. I thought it very sociable in a stranger. Presently, in the course of conversation, I made a statement bearing upon the subject under discussion — and I made it with due modesty, for there was nothing extraordinary about it, and it was only put forth in illustration of a point at issue. I had barely finished when this person spoke out with rapid utterance and feverish anxiety:

"Oh, that was certainly remarkable, after a fashion, but you ought to have seen *my* chimney — you ought to have seen *my* chimney, sir! Smoke! I wish I may hang if — Mr. Jones, *you* remember that chimney — you *must* remember that chimney!

(317)

No, no — I recollect, now, you warn't living on this
side of the island then. But I am telling you nothing but the truth, and I wish I may never draw
another breath if that chimney didn't smoke so that
the smoke actually got *caked* in it and I had to dig
it out with a pickaxe! You may smile, gentlemen,
but the high sheriff's got a hunk of it which I dug
out before his eyes, and so it's perfectly easy for
you to go and examine for yourselves."

The interruption broke up the conversation, which
had already begun to lag, and we presently hired
some natives and an outrigger canoe or two, and
went out to overlook a grand surf-bathing contest.

Two weeks after this, while talking in a company,
I looked up and detected this same man boring
through and through me with his intense eye, and
noted again his twitching muscles and his feverish
anxiety to speak. The moment I paused, he said:

" *Beg* your pardon, sir, beg your pardon, but it
can only be considered remarkable when brought
into strong outline by isolation. Sir, contrasted
with a circumstance which occurred in my own experience, it instantly becomes commonplace. No,
not that — for I will not speak so discourteously of
any experience in the career of a stranger and a
gentleman — but I am *obliged* to say that you could
not, and you *would* not ever again refer to this tree
as a *large* one, if you could behold, as I have, the
great Yakmatack tree, in the island of Ounaska, sea
of Kamtchatka — a tree, sir, not one inch less than

four hundred and fifteen feet in solid diameter!—
and I wish I may die in a minute if it isn't so! Oh,
you needn't look so questioning, gentlemen; here's
old Cap Saltmarsh can say whether I know what
I'm talking about or not. I showed him the
tree."

Captain Saltmarsh.—"Come, now, cat your
anchor, lad — you're heaving too taut. You *promised* to show me that stunner, and I walked more
than eleven mile with you through the cussedest
jungle *I* ever see, a hunting for it; but the tree you
showed me finally warn't as big around as a beer
cask, and *you* know that your own self, Markiss."

"Hear the man talk! Of *course* the tree was reduced that way, but didn't I *explain* it? Answer
me, didn't I? Didn't I say I wished you could
have seen it when *I* first saw it? When you got up
on your ear and called me names, and said I had
brought you eleven miles to look at a sapling, didn't
I *explain* to you that all the whaleships in the
North Seas had been wooding off of it for more
than twenty-seven years? And did you s'pose the
tree could last for-*ever*, con-*found* it? I don't see
why you want to keep back things that way, and try
to injure a person that's never done *you* any harm."

Somehow this man's presence made me uncom-
fortable, and I was glad when a native arrived at
that moment to say that Muckawow, the most com-
panionable and luxurious among the rude war-chiefs
of the Islands, desired us to come over and help

him enjoy a missionary whom he had found trespassing on his grounds.

I think it was about ten days afterward that, as I finished a statement I was making for the instruction of a group of friends and acquaintances, and which made no pretense of being extraordinary, a familiar voice chimed instantly in on the heels of my last word, and said:

"But, my dear sir, there was *nothing* remarkable about that horse, or the circumstance either — nothing in the world! I mean no sort of offense when I say it, sir, but you really do not know anything whatever about speed. Bless your heart, if you could only have seen my mare Margaretta; *there* was a beast! — *there* was lightning for you! Trot! Trot is no name for it — she flew! How she *could* whirl a buggy along! I started her out once, sir — Colonel Bilgewater, *you* recollect that animal perfectly well — I started her out about thirty or thirty-five yards ahead of the awfullest storm I ever saw in my life, and it chased us upwards of eighteen miles! It did, by the everlasting hills! And I'm telling you nothing but the unvarnished truth when I say that not one single drop of rain fell on me — not a single *drop*, sir! And I swear to it! But my dog was a-swimming behind the wagon all the way!"

For a week or two I stayed mostly within doors, for I seemed to meet this person everywhere, and he had become utterly hateful to me. But one evening

I dropped in on Captain Perkins and his friends, and we had a sociable time. About ten o'clock I chanced to be talking about a merchant friend of mine, and without really intending it, the remark slipped out that he was a little mean and parsimonious about paying his workmen. Instantly, through the steam of a hot whisky punch on the opposite side of the room, a remembered voice shot— and for a moment I trembled on the imminent verge of profanity:

"Oh, my dear sir, really you expose yourself when you parade *that* as a surprising circumstance. Bless your heart and hide, you are ignorant of the very A B C of meanness! ignorant as the unborn babe! ignorant as unborn *twins!* You don't know *anything* about it! It is pitiable to see you, sir, a well-spoken and prepossessing stranger, making such an enormous pow-wow here about a subject concerning which your ignorance is perfectly humiliating! Look me in the eye, if you please; look me in the eye. John James Godfrey was the son of poor but honest parents in the State of Mississippi — boyhood friend of mine — bosom comrade in later years. Heaven rest his noble spirit, he is gone from us now. John James Godfrey was hired by the Hayblossom Mining Company in California to do some blasting for them — the 'Incorporated Company of Mean Men,' the boys used to call it. Well, one day he drilled a hole about four feet deep and put in an awful blast of powder, and was stand-

21**

ing over it ramming it down with an iron crowbar
about nine foot long, when the cussed thing struck
a spark and fired the powder, and scat! away John
Godfrey whizzed like a sky-rocket, him and his crowbar! Well, sir, he kept on going up in the air
higher and higher, till he didn't look any bigger
than a boy — and he kept going on up higher and
higher, till he didn't look any bigger than a doll —
and he kept on going up higher and higher, till he
didn't look any bigger than a little small bee — and
then he went out of sight! Presently he came in
sight again, looking like a little small bee — and he
came along down further and further, till he looked
as big as a doll again — and down further and
further, till he was as big as a boy again — and
further and further, till he was a full-sized man once
more; and then him and his crowbar came a
wh-izzing down and lit right exactly in the same old
tracks and went to r-ramming down, and r-ramming
down, and r-ramming down again, just the same as
if nothing had happened! Now, do you know, that
poor cuss warn't gone only sixteen minutes, and yet
that incorporated company of mean men DOCKED
HIM FOR THE LOST TIME!"

I said I had the headache, and so excused myself
and went home. And on my diary I entered
"another night spoiled" by this offensive loafer.
And a fervent curse was set down with it to keep
the item company. And the very next day I packed
up, out of all patience, and left the island.

Almost from the very beginning, I regarded that man as a liar.

. . . .

The line of points represents an interval of years. At the end of which time the opinion hazarded in that last sentence came to be gratifyingly and remarkably endorsed, and by wholly disinterested persons. The man Markiss was found one morning hanging to a beam of his own bedroom (the doors and windows securely fastened on the inside), dead · and on his breast was pinned a paper in his own handwriting begging his friends to suspect no inno-cent person of having anything to do with his death, for that it was the work of his own hands entirely. Yet the jury brought in the astounding verdict that deceased came to his death " by the hands of some person or persons unknown!" They explained that the perfectly undeviating consistency of Markiss's character for thirty years towered aloft as colossal and indestructible testimony, that whatever state-ment he chose to make was entitled to instant and unquestioning acceptance as a *lie*. And they furthermore stated their belief that he was not dead, and instanced the strong circumstantial evi-dence of his own word that he *was* dead — and beseeched the coroner to delay the funeral as long as possible, which was done. And so in the tropical climate of Lahaina the coffin stood open for seven days, and then even the loyal jury gave him up. But they sat on him again, and changed their verdict

v**

to "suicide induced by mental aberration"—because, said they, with penetration, "he said he was
dead, and he *was* dead; and would he have told the
truth if he had been in his right mind? *No,* sir."

CHAPTER XXXVII.

AFTER half a year's luxurious vagrancy in the Islands, I took shipping in a sailing vessel, and regretfully returned to San Francisco — a voyage in every way delightful, but without an incident; unless lying two long weeks in a dead calm, eighteen hundred miles from the nearest land, may rank as an incident. Schools of whales grew so tame that day after day they played about the ship among the porpoises and the sharks without the least apparent fear of us, and we pelted them with empty bottles for lack of better sport. Twenty-four hours afterward these bottles would be still lying on the glassy water under our noses, showing that the ship had not moved out of her place in all that time. The calm was absolutely breathless, and the surface of the sea absolutely without a wrinkle. For a whole day and part of a night we lay so close to another ship that had drifted to our vicinity, that we carried on conversations with her passengers, introduced each other by name, and became pretty intimately acquainted with people we had never heard of before, and have never heard of since. This was the

11   (325)

only vessel we saw during the whole lonely voyage. We had fifteen passengers, and to show how hard pressed they were at last for occupation and amusement, I will mention that the gentlemen gave a good part of their time every day, during the calm, to trying to sit on an empty champagne bottle (lying on its side) and thread a needle without touching their heels to the deck, or falling over; and the ladies sat in the shade of the mainsail, and watched the enterprise with absorbing interest. We were at sea five Sundays; and yet, but for the almanac, we never would have known but that all the other days were Sundays too.

I was home again, in San Francisco, without means and without employment. I tortured my brain for a saving scheme of some kind, and at last a public lecture occurred to me! I sat down and wrote one, in a fever of hopeful anticipation. I showed it to several friends, but they all shook their heads. They said nobody would come to hear me, and I would make a humiliating failure of it. They said that as I had never spoken in public, I would break down in the delivery, anyhow. I was disconsolate now. But at last an editor slapped me on the back and told me to "go ahead." He said, "Take the largest house in town, and charge a dollar a ticket." The audacity of the proposition was charming; it seemed fraught with practical worldly wisdom, however. The proprietor of the several theaters endorsed the advice, and said I might have

his handsome new opera house at half price — fifty
dollars. In sheer desperation I took it — on credit,
for sufficient reasons. In three days I did a hundred and fifty dollars' worth of printing and advertising, and was the most distressed and frightened
creature on the Pacific coast. I could not sleep —
who could, under such circumstances? For other
people there was facetiousness in the last line of my
posters, but to me it was plaintive with a pang when
I wrote it:

"Doors open at 7½. The trouble will begin at 8."

That line has done good service since. Showmen
have borrowed it frequently. I have even seen it
appended to a newspaper advertisement reminding
school pupils in vacation what time next term would
begin. As those three days of suspense dragged
by, I grew more and more unhappy. I had sold
two hundred tickets among my personal friends, but
I feared they might not come. My lecture, which
had seemed "humorous" to me, at first, grew
steadily more and more dreary, till not a vestige of
fun seemed left, and I grieved that I could not bring
a coffin on the stage and turn the thing into a
funeral. I was so panic-stricken, at last, that I
went to three old friends, giants in stature, cordial
by nature, and stormy-voiced, and said:

"This thing is going to be a failure; the jokes in
it are so dim that nobody will ever see them; I
would like to have you sit in the parquette, and help
me through."

They said they would. Then I went to the wife of a popular citizen, and said that if she was willing to do me a very great kindness, I would be glad if she and her husband would sit prominently in the left-hand stage-box, where the whole house could see them. I explained that I should need help, and would turn toward her and smile, as a signal, when I had been delivered of an obscure joke — "and then," I added, "don't wait to investigate, but *respond!*"

She promised. Down the street I met a man I never had seen before. He had been drinking, and was beaming with smiles and good nature. He said:

"My name's Sawyer. You don't know me, but that don't matter. I haven't got a cent, but if you knew how bad I wanted to laugh, you'd give me a ticket. Come, now, what do you say?"

"Is your laugh hung on a hair-trigger? — that is, is it critical, or can you get it off *easy?*"

My drawling infirmity of speech so affected him that he laughed a specimen or two that struck me as being about the article I wanted, and I gave him a ticket, and appointed him to sit in the second circle, in the center, and be responsible for that division of the house. I gave him minute instructions about how to detect indistinct jokes, and then went away, and left him chuckling placidly over the novelty of the idea.

I ate nothing on the last of the three eventful days

—I only suffered. I had advertised that on this
third day the box-office would be opened for the
sale of reserved seats. I crept down to the theater
at four in the afternoon to see if any sales had been
made. The ticket-seller was gone, the box-office
was locked up. I had to swallow suddenly, or my
heart would have got out. "No sales," I said to
myself; "I might have known it." I thought of
suicide, pretended illness, flight. I thought of these
things in earnest, for I was very miserable and
scared. But of course I had to drive them away,
and prepare to meet my fate. I could not wait for
half-past seven — I wanted to face the horror, and
end it — the feeling of many a man doomed to hang,
no doubt. I went down back streets at six o'clock,
and entered the theater by the back door. I stumbled my way in the dark among the ranks of canvas
scenery, and stood on the stage. The house was
gloomy and silent, and its emptiness depressing. I
went into the dark among the scenes again, and for
an hour and a half gave myself up to the horrors,
wholly unconscious of everything else. Then I
heard a murmur; it rose higher and higher, and
ended in a crash, mingled with cheers. It made my
hair raise, it was so close to me, and so loud. There
was a pause, and then another; presently came a
third, and before I well knew what I was about, I
was in the middle of the stage, staring at a sea of
faces, bewildered by the fierce glare of the lights,
and quaking in every limb with a terror that seemed

like to take my life away. The house was full,
aisles and all!

The tumult in my heart and brain and legs continued a full minute before I could gain any command over myself. Then I recognized the charity
and the friendliness in the faces before me, and little
by little my fright melted away, and I began to talk.
Within three or four minutes I was comfortable, and
even content. My three chief allies, with three
auxiliaries, were on hand, in the parquette, all sitting together, all armed with bludgeons, and all
ready to make an onslaught upon the feeblest joke
that might show its head. And whenever a joke
did fall, their bludgeons came down and their faces
seemed to split from ear to ear; Sawyer, whose
hearty countenance was seen looming redly in the
center of the second circle, took it up, and the house
was carried handsomely. Inferior jokes never fared
so royally before. Presently, I delivered a bit of
serious matter with impressive unction (it was my
pet), and the audience listened with an absorbed
hush that gratified me more than any applause; and
as I dropped the last word of the clause, I happened
to turn and catch Mrs. ——'s intent and waiting
eye; my conversation with her flashed upon me,
and in spite of all I could do I smiled. She took it
for the signal, and promptly delivered a mellow
laugh that touched off the whole audience; and the
explosion that followed was the triumph of the
evening. I thought that that honest man Sawyer

would choke himself; and as for the bludgeons, they performed like pile-drivers. But my poor little morsel of pathos was ruined. It was taken in good faith as an intentional joke, and the prize one of the entertainment, and I wisely let it go at that.

All the papers were kind in the morning; my appetite returned; I had abundance of money. All's well that ends well.

CHAPTER XXXVIII.

I LAUNCHED out as a lecturer, now, with great boldness. I had the field all to myself, for public lectures were almost an unknown commodity in the Pacific market. They are not so rare, now, I suppose. I took an old personal friend along to play agent for me, and for two or three weeks we roamed through Nevada and California and had a very cheerful time of it. Two days before I lectured in Virginia City, two stage-coaches were robbed within two miles of the town. The daring act was committed just at dawn, by six masked men, who sprang up alongside the coaches, presented revolvers at the heads of the drivers and passengers, and commanded a general dismount. Everybody climbed down, and the robbers took their watches and every cent they had. Then they took gunpowder and blew up the express specie boxes and got their contents. The leader of the robbers was a small, quick-spoken man, and the fame of his vigorous manner and his intrepidity was in everybody's mouth when we arrived.

The night after instructing Virginia, I walked over

the desolate " divide " and down to Gold Hill, and
lectured there. The lecture done, I stopped to talk
with a friend, and did not start back till eleven.
The " divide " was high, unoccupied ground, between the towns, the scene of twenty midnight murders and a hundred robberies. As we climbed up
and stepped out on this eminence, the Gold Hill
lights dropped out of sight at our backs, and the
night closed down gloomy and dismal. A sharp
wind swept the place, too, and chilled our perspiring
bodies through.

"I tell you I don't like this place at night," said
Mike, the agent.

"Well, don't speak so loud," I said. "You
needn't remind anybody that we are here."

Just then a dim figure approached me from the
direction of Virginia — a man, evidently. He came
straight at me, and I stepped aside to let him pass;
he stepped in the way and confronted me again.
Then I saw that he had a mask on and was holding
something in my face — I heard a click-click and
recognized a revolver in dim outline. I pushed the
barrel aside with my hand and said:

"Don't!"

He ejaculated sharply:

"Your watch! Your money!"

I said.

"You can have them with pleasure — but take
the pistol away from my face, please. It makes me
shiver."

"No remarks! Hand out your money!"

"Certainly — I —"

"Put up your hands! Don't you go for a weapon! Put 'em up! Higher!"

I held them above my head.

A pause. Then:

"Are you going to hand out your money or not?"

I dropped my hands to my pockets and said:

"Certainly! I —"

"Put up your *hands!* Do you want your head blown off? Higher!"

I put them above my head again.

Another pause.

"*Are* you going to hand out your money or *not?* Ah-ah — again? Put up your hands! By George, you want the head shot off you awful bad!"

"Well, friend, I'm trying my best to please you. You tell me to give up my money, and when I reach for it you tell me to put up my hands. If you would only —. Oh, now — don't! All six of you at me! That other man will get away while — Now, please take some of those revolvers out of my face — *do*, if you *please!* Every time one of them clicks, my liver comes up into my throat! If you have a mother — any of you — or if any of you have ever *had* a mother — or a — grandmother — or a —"

"Cheese it! *Will* you give up your money, or have we got to — There — there — none of that! Put up your *hands!*"

"Gentlemen — I know you are gentlemen by your —"

"Silence! If you want to be facetious, young man, there are times and places more fitting. *This* is a serious business."

"You prick the marrow of my opinion. The funerals I have attended in my time were comedies compared to it. Now, *I* think —"

"Curse your palaver! Your money! — your money! — your money! Hold! — put up your hands!"

"Gentlemen, listen to reason. You *see* how I am situated — now *don't* put those pistols so close — I smell the powder. You see how I am situated. If I had four hands — so that I could hold up two and —"

"Throttle him! Gag him! Kill him!"

"Gentlemen, *don't!* Nobody's watching the other fellow. Why don't some of you — Ouch! Take it away, please! Gentlemen, you see that I've got to hold up my hands; and so I can't take out my money — but if you'll be so kind as to take it out for me, I will do as much for you some —"

"Search him, Beauregard — and stop his jaw with a bullet, quick, if he wags it again. Help, Beauregard, Stonewall."

Then three of them, with the small, spry leader, adjourned to Mike and fell to searching him. I was so excited that my lawless fancy tortured me to ask my two men all manner of facetious questions about their rebel brother-generals of the South, but, con-

sidering the order they had received, it was but
common prudence to keep still. When everything
had been taken from me,— watch, money, and a
multitude of trifles of small value,— I supposed I
was free, and forthwith put my cold hands into my
empty pockets and began an inoffensive jig to warm
my feet and stir up some latent courage — but instantly all pistols were at my head, and the order
came again:

" Be still! Put up your hands! And *keep* them
up!"

They stood Mike up alongside of me, with strict
orders to keep his hands above his head, too, and
then the chief highwayman said:

" Beauregard, hide behind that boulder; Phil
Sheridan, you hide behind that other one; Stonewall Jackson, put yourself behind that sage-bush
there. Keep your pistols bearing on these fellows,
and if they take down their hands within ten minutes, or move a single peg, let them have it!"

Then three disappeared in the gloom toward the
several ambushes, and the other three disappeared
down the road toward Virginia.

It was depressingly still, and miserably cold.
Now, this whole thing was a practical joke, and the
robbers were personal friends of ours in disguise,
and twenty more lay hidden within ten feet of us
during the whole operation, listening. Mike knew all
of this, and was in the joke, but I suspected nothing
of it. To me it was most uncomfortably genuine.

When we had stood there in the middle of the road five minutes, like a couple of idiots, with our hands aloft, freezing to death by inches, Mike's interest in the joke began to wane. He said:

" The time's up, now, ain't it?"

" No, you keep still. Do you want to take any chances with those bloody savages?"

Presently Mike said:

" *Now* the time's up, anyway. I'm freezing."

" Well, freeze. Better freeze than carry your brains home in a basket. Maybe the time *is* up, but how do *we* know?— got no watch to tell by. I mean to give them good measure. I calculate to stand here fifteen minutes or die. Don't you move."

So, without knowing it, I was making one joker very sick of his contract. When we took our arms down at last, they were aching with cold and fatigue, and when we went sneaking off, the dread I was in that the time might not yet be up and that we would feel bullets in a moment, was not sufficient to draw all my attention from the misery that racked my stiffened body.

The joke of these highwayman friends of ours was mainly a joke upon themselves; for they had waited for me on the cold hilltop two full hours before I came, and there was very little fun in that; they were so chilled that it took them a couple of weeks to get warm again. Moreover, I never had a thought that they would kill me to get money which

22**

it was so perfectly easy to get without any such
folly, and so they did not really frighten me bad
enough to make their enjoyment worth the trouble
they had taken. I was only afraid that their weapons
would go off accidentally. Their very numbers inspired me with confidence that no blood would be
intentionally spilled. They were not smart; they
ought to have sent only *one* highwayman, with a
double-barreled shotgun, if they desired to see the
author of this volume climb a tree.

However, I suppose that in the long run I got the
largest share of the joke at last; and in a shape not
foreseen by the highwaymen; for the chilly exposure on the "divide" while I was in a perspiration
gave me a cold which developed itself into a troublesome disease and kept my hands idle some three
months, besides costing me quite a sum in doctor's
bills. Since then I play no practical jokes on people and generally lose my temper when one is played
upon me.

When I returned to San Francisco I projected a
pleasure journey to Japan and thence westward
around the world; but a desire to see home again
changed my mind, and I took a berth in the steam-
ship, bade good-bye to the friendliest land and
livest, heartiest community on our continent, and
came by the way of the Isthmus to New York — a
trip that was not much of a picnic excursion, for
the cholera broke out among us on the passage, and
we buried two or three bodies at sea every day. I

found home a dreary place after my long absence; for half the children I had known were now wearing whiskers or waterfalls, and few of the grown people I had been acquainted with remained at their hearthstones prosperous and happy — some of them had wandered to other scenes, some were in jail, and the rest had been hanged. These changes touched me deeply, and I went away and joined the famous Quaker City European Excursion and carried my tears to foreign lands.

Thus, after seven years of vicissitudes, ended a "pleasure trip" to the silver mines of Nevada which had originally been intended to occupy only three months. However, I usually miss my calculations further than that.

MORAL.

If the reader thinks he is done, now, and that this book has no moral to it, he is in error. The moral of it is this: If you are of any account, stay at home and make your way by faithful diligence; but if you are "no account," go away from home, and then you will *have* to work, whether you want to or not. Thus you become a blessing to your friends by ceasing to be a nuisance to them — if the people you go among suffer by the operation.

APPENDIX

APPENDIX

A

BRIEF SKETCH OF MORMON HISTORY.

MORMONISM is only about forty years old, but its career has been full of stir and adventure from the beginning, and is likely to remain so to the end. Its adherents have been hunted and hounded from one end of the country to the other, and the result is that for years they have hated all "Gentiles" indiscriminately and with all their might. Joseph Smith, the finder of the Book of Mormon and founder of the religion, was driven from State to State with his mysterious copperplates and the miraculous stones he read their inscriptions with. Finally he instituted his "church" in Ohio and Brigham Young joined it. The neighbors began to persecute, and apostasy commenced. Brigham held to the faith and worked hard. He arrested desertion. He did more — he added converts in the midst of the trouble. He rose in favor and importance with the brethren. He was made one of the Twelve Apostles of the Church. He shortly fought his way to a higher post and a more powerful — President of the Twelve. The neighbors rose up and drove the Mormons out of Ohio, and they settled in Missouri. Brigham went with them. The Missourians drove them out, and they retreated to Nauvoo, Illinois. They prospered there, and built a temple which made some pretensions to architectural grace and achieved some celebrity in a section of country where a brick court-house with a tin dome and a cupola on it was contemplated with reverential awe. But the Mormons were badgered and harried again by their neighbors. All the proclamations Joseph Smith could issue denouncing polygamy and repudiating it as utterly anti-Mormon were of no avail; the people of the

(343)

neighborhood, on both sides of the Mississippi, claimed that polygamy was practiced by the Mormons, and not only polygamy but a little of everything that was bad. Brigham returned from a mission to England, where he had established a Mormon newspaper, and he brought back with him several hundred converts to his preaching. His influence among the brethren augmented with every move he made. Finally, Nauvoo was invaded by the Missouri and Illinois Gentiles, and Joseph Smith killed. A Mormon named Rigdon assumed the presidency of the Mormon church and government, in Smith's place, and even tried his hand at a prophecy or two. But a greater than he was at hand. Brigham seized the advantage of the hour and without other authority than superior brain and nerve and will, hurled Rigdon from his high place and occupied it himself. He did more. He launched an elaborate curse at Rigdon and his disciples; and he pronounced Rigdon's " prophecies " emanations from the devil, and ended by " handing the false prophet over to the buffetings of Satan for a thousand years " — probably the longest term ever inflicted in Illinois. The people recognized their master. They straightway elected Brigham Young President, by a prodigious majority, and have never faltered in their devotion to him from that day to this. Brigham had forecast — a quality which no other prominent Mormon has probably ever possessed. He recognized that it was better to move to the wilderness than be moved. By his command the people gathered together their meager effects, turned their backs upon their homes, and their faces toward the wilderness, and on a bitter night in February filed in sorrowful procession across the frozen Mississippi, lighted on their way by the glare from their burning temple, whose sacred furniture their own hands had fired! They camped, several days afterward, on the western verge of Iowa, and poverty, want, hunger, cold, sickness, grief, and persecution did their work, and many succumbed and died — martyrs, fair and true, whatever else they might have been. Two years the remnant remained there, while Brigham and a small party crossed the country and founded Great Salt Lake City, purposely choosing a land which was *outside the ownership and jurisdiction of the hated American nation.* Note that. This was in 1847. Brigham moved his people there and got them settled just in time to see disaster fall again. For the war closed and Mexico ceded Brigham's refuge to the enemy — the United States! In 1849 the Mormons organized a " free and independent " government and erected the " State of Deseret," with Brigham Young as its head. But

the very next year Congress deliberately snubbed it and created the
" Territory of Utah " out of the same accumulation of mountains, sage-
brush, alkali, and general desolation,— but made Brigham Governor of
it. Then for years the enormous migration across the plains to California poured through the land of the Mormons, and yet the church
remained staunch and true to its lord and master. Neither hunger,
thirst, poverty, grief, hatred, contempt, nor persecution could drive the
Mormons from their faith or their allegiance; and even the thirst for
gold, which gleaned the flower of the youth and strength of many
nations, was not able to entice them! That was the final test. An
experiment that could survive that was an experiment with some substance to it somewhere.

Great Salt Lake City throve finely, and so did Utah. One of the
last things which Brigham Young had done before leaving Iowa, was to
appear in the pulpit dressed to personate the worshiped and lamented
prophet Smith, and confer the prophetic succession, with all its digni-
ties, emoluments, and authorities, upon " President Brigham Young " !
The people accepted the pious fraud with the maddest enthusiasm, and
Brigham's power was sealed and secured for all time. Within five years
afterward he openly added polygamy to the tenets of the church by
authority of a " revelation " which he pretended had been received nine
years before by Joseph Smith, albeit Joseph is amply on record as
denouncing polygamy to the day of his death.

Now was Brigham become a second Andrew Johnson in the small
beginning and steady progress in his official grandeur. He had served
successively as a disciple in the ranks; home missionary; foreign missionary; editor and publisher; Apostle; President of the Board of
Apostles; President of all Mormondom, civil and ecclesiastical; successor to the great Joseph by the will of heaven; "prophet," " seer,"
" revelator." There was but one dignity higher which he *could* aspire
to, and he reached out modestly and took that — he proclaimed himself
a God!

He claims that he is to have a heaven of his own hereafter, and that
he will be its God, and his wives and children its goddesses, princes and
princesses. Into it all faithful Mormons will be admitted, with their
families, and will take rank and consequence according to the number
of their wives and children. If a disciple dies before he has had time to
accumulate enough wives and children to enable him to be respectable
in the next world any friend can marry a few wives and raise a few

children for him *after he is dead*, and they are duly credited to his account and his heavenly status advanced accordingly.

Let it be borne in mind that the majority of the Mormons have always been ignorant, simple, of an inferior order of intellect, unacquainted with the world and its ways; and let it be borne in mind that the wives of these Mormons are necessarily after the same pattern, and their children likely to be fit representatives of such a conjunction; and then let it be remembered that *for forty years* these creatures have been driven, driven, driven, relentlessly! and mobbed, beaten, and shot down; cursed, despised, expatriated; banished to a remote desert, whither they journeyed gaunt with famine and disease, disturbing the ancient solitudes with their lamentations and marking the long way with graves of their dead — and all because they were simply trying to live and worship God in the way which *they* believed with all their hearts and souls to be the true one. Let all these things be borne in mind, and then it will not be hard to account for the deathless hatred which the Mormons bear our people and our government.

That hatred has " fed fat its ancient grudge " ever since Mormon Utah developed into a self-supporting realm and the church waxed rich and strong. Brigham as Territorial Governor made it plain that Mormondom was for the Mormons. The United States tried to rectify all that by appointing territorial officers from New England and other anti-Mormon localities, but Brigham prepared to make their entrance into his dominions difficult. Three thousand United States troops had to go across the plains and put these gentlemen in office. And after they were in office they were as helpless as so many stone images. They made laws which nobody minded and which could not be executed. The federal judges opened court in a land filled with crime and violence and sat as holiday spectacles for insolent crowds to gape at — for there was nothing to try, nothing to do, nothing on the dockets! And if a Gentile brought a suit, the Mormon jury would do just as it pleased about bringing in a verdict, and when the judgment of the court was rendered no Mormon cared for it and no officer could execute it. Our Presidents shipped one cargo of officials after another to Utah, but the result was always the same — they sat in a blight for awhile, they fairly feasted on scowls and insults day by day, they saw every attempt to do their official duties find its reward in darker and darker looks, and in secret threats and warnings of a more and more dismal nature — and at last they either succumbed and became despised tools and toys of the

Mormons, or got scared and discomforted beyond all endurance and left the territory. If a brave officer kept on courageously till his pluck was proven, some pliant Buchanan or Pierce would remove him and appoint a stick in his place. In 1857 General Harney came very near being appointed Governor of Utah. And so it came very near being Harney governor and Cradlebaugh judge! — two men who never had any idea of fear further than the sort of murky comprehension of it which they were enabled to gather from the dictionary. Simply (if for nothing else) for the variety they would have made in a rather monotonous history of Federal servility and helplessness, it is a pity they were not fated to hold office together in Utah.

Up to the date of our visit to Utah, such had been the Territorial record. The Territorial government established there had been a hopeless failure, and Brigham Young was the only real power in the land. He was an absolute monarch — a monarch who defied our President — a monarch who laughed at our armies when they camped about his capital — a monarch who received without emotion the news that the august Congress of the United States had enacted a solemn law against polygamy, and then went forth calmly and married twenty-five or thirty more wives.

B

THE MOUNTAIN MEADOWS MASSACRE.

THE persecutions which the Mormons suffered so long — and which they consider they still suffer in not being allowed to govern themselves — they have endeavored and are still endeavoring to repay. The now almost forgotten "Mountain Meadows massacre" was their work. It was very famous in its day. The whole United States rang with its horrors. A few items will refresh the reader's memory. A great emigrant train from Missouri and Arkansas passed through Salt Lake City, and a few disaffected Mormons joined it for the sake of the strong protection it afforded for their escape. In that matter lay sufficient cause for hot retaliation by the Mormon chiefs. Besides, these one hundred and forty-five or one hundred and fifty unsuspecting emigrants being in part from Arkansas, where a noted Mormon missionary had lately been killed, and in part from Missouri, a State remembered with execrations as a bitter persecutor of the saints when they were few and poor and friendless, here were substantial additional grounds for lack of love for these wayfarers. And finally, this train was rich, very rich in cattle, horses, mules, and other property — and how could the Mormons consistently keep up their coveted resemblance to the Israelitish tribes and not seize the "spoil" of an enemy when the Lord had so manifestly "delivered it into their hand"?

Wherefore, according to Mrs. C. V. Waite's entertaining book, "The Mormon Prophet," it transpired that —

"A 'revelation' from Brigham Young, as Great Grand Archee or God, was despatched to President J. C. Haight, Bishop Higbee, and J. D. Lee (adopted son of Brigham), commanding them to raise all the forces they could muster and trust, follow those cursed Gentiles (so read the revelation), attack them disguised as Indians, and with the arrows of the Almighty make a clean sweep of them, and leave none to tell the tale; and if they needed any assistance they were commanded to hire

(348)

the Indians as their allies, promising them a share of the booty. They were to be neither slothful nor negligent in their duty, and to be punctual in sending the teams back to him before winter set in, for this was the mandate of Almighty God."

The command of the "revelation" was faithfully obeyed. A large party of Mormons, painted and tricked out as Indians, overtook the train of emigrant wagons some three hundred miles south of Salt Lake City, and made an attack. But the emigrants threw up earthworks, made fortresses of their wagons, and defended themselves gallantly and successfully for five days! Your Missouri or Arkansas gentleman is not much afraid of the sort of scurvy apologies for "Indians" which the southern part of Utah affords. He would stand up and fight five hundred of them.

At the end of the five days the Mormons tried military strategy. They retired to the upper end of the "Meadows," resumed civilized apparel, washed off their paint, and then, heavily armed, drove down in wagons to the beleaguered emigrants, bearing a flag of truce! When the emigrants saw white men coming they threw down their guns and welcomed them with cheer after cheer! And, all unconscious of the poetry of it, no doubt, they lifted a little child aloft, dressed in white, in answer to the flag of truce!

The leaders of the timely white "deliverers" were President Haight and Bishop John D. Lee, of the Mormon Church. Mr. Cradlebaugh, who served a term as a Federal Judge in Utah and afterward was sent to Congress from Nevada, tells in a speech delivered in Congress how these leaders next proceeded:

"They professed to be on good terms with the Indians, and represented them as being very mad. They also proposed to intercede and settle the matter with the Indians. After several hours' parley they, having (apparently) visited the Indians, gave the *ultimatum* of the savages; which was, that the emigrants should march out of their camp, leaving everything behind them, even their guns. It was promised by the Mormon bishops that they would bring a force and guard the emigrants back to the settlements. The terms were agreed to, the emigrants being desirous of saving the lives of their families. The Mormons retired, and subsequently appeared with thirty or forty armed men. The emigrants were marched out, the women and children in front and the men behind, the Mormon guard being in the rear. When they had marched in this way about a mile, at a given signal the slaughter commenced. The men were almost all shot down at the first fire from the guard. Two only escaped, who fled to the desert, and were followed one hundred and fifty miles before they were overtaken and slaughtered. The women and children ran on, two or three hundred yards further, when they were overtaken and with the aid of the Indians they were slaughtered. Seventeen individuals only, of all the emigrant party, were spared, and they were little children, the eldest of them being only seven years old. Thus, on the 10th day of September, 1857, was consummated one of the most cruel, cowardly, and bloody murders known in our history."

The number of persons butchered by the Mormons on this occasion was *one hundred and twenty.*

With unheard-of temerity Judge Cradlebaugh opened his court and proceeded to make Mormondom answer for the massacre. And what a spectacle it must have been to see this grim veteran, solitary and alone in his pride and his pluck, glowering down on his Mormon jury and Mormon auditory, deriding them by turns, and by turns "breathing threatenings and slaughter"!

An editorial in the *Territorial Enterprise* of that day says of him and of the occasion:

"He spoke and acted with the fearlessness and resolution of a Jackson; but the jury failed to indict, or even report on the charges, while threats of violence were heard in every quarter, and an attack on the U. S. troops intimated, if he persisted in his course.

"Finding that nothing could be done with the juries, they were discharged, with a scathing rebuke from the judge. And then, sitting as a committing magistrate, *he commenced his task alone.* He examined witnesses, made arrests in every quarter, and created a consternation in the camps of the saints greater than any they had ever witnessed before, since Mormondom was born. At last accounts terrified elders and bishops were decamping to save their necks; and developments of the most startling character were being made, implicating the highest Church dignitaries in the many murders and robberies committed upon the Gentiles during the past eight years."

Had Harney been Governor, Cradlebaugh would have been supported in his work, and the absolute proofs adduced by him of Mormon guilt in this massacre and in a number of previous murders, would have conferred gratuitous coffins upon certain citizens, together with occasion to use them. But Cumming was the Federal Governor, and he, under a curious pretense of impartiality, sought to screen the Mormons from the demands of justice. On one occasion he even went so far as to publish his protest against the use of the U. S. troops in aid of Cradlebaugh's proceedings.

Mrs. C. V. Waite closes her interesting detail of the great massacre with the following remark and accompanying summary of the testimony — and the summary is concise, accurate, and reliable:

"For the benefit of those who may still be disposed to doubt the guilt of Young and his Mormons in this transaction, the testimony is here collated and circumstances given which go not merely to implicate but to fasten conviction upon them by 'confirmations strong as proofs of Holy Writ':

"1. The evidence of Mormons themselves, engaged in the affair, as shown by the statements of Judge Cradlebaugh and Deputy U. S. Marshal Rodgers.

"2. The failure of Brigham Young to embody any account of it in his Report as Superintendent of Indian Affairs. Also his failure to make any allusion to it whatever from the pulpit, until several years after the occurrence.

"3. The flight to the mountains of men high in authority in the Mormon Church and State, when this affair was brought to the ordeal of a judicial investigation.

"4. The failure of the *Deseret News*, the Church organ, and the only paper then published in the Territory, to notice the massacre until several months afterward, and then only to deny that Mormons were engaged in it.

"5. The testimony of the children saved from the massacre.

"6. The children and the property of the emigrants found in possession of the Mormons, and that possession traced back to the very day after the massacre.

"7. The statements of Indians in the neighborhood of the scene of the massacre; these statements are shown, not only by Cradlebaugh and Rodgers, but by a number of military officers, and by J. Forney, who was, in 1859, Superintendent of Indian Affairs for the Territory. To all these were such statements freely and frequently made by the Indians.

"8. The testimony of R. P. Campbell, Capt. 2d Dragoons, who was sent in the spring of 1859 to Santa Clara, to protect travelers on the road to California and to inquire into Indian depredations."

23

C

CONCERNING A FRIGHTFUL ASSASSINATION THAT WAS NEVER CONSUMMATED.

[IF ever there was a harmless man, it is Conrad Wiegand, of Gold Hill, Nevada. If ever there was a gentle spirit that thought itself un-fired gunpowder and latent ruin, it is Conrad Wiegand. If ever there was an oyster that fancied itself a whale; or a jack-o'lantern, confined to a swamp, that fancied itself a planet with a billion-mile orbit; or a summer zephyr that deemed itself a hurricane, it is Conrad Wiegand. Therefore, what wonder is it that when he says a thing, he thinks the world listens; that when he does a thing the world stands still to look; and that when he suffers, there is a convulsion of nature? When I met Conrad, he was "Superintendent of the Gold Hill Assay Office"—and he was not only its Superintendent, but its entire force. And he was a street preacher, too, with a mongrel religion of his own inven-tion, whereby he expected to regenerate the universe. This was years ago. Here latterly he has entered journalism; and his journalism is what it might be expected to be: colossal to ear, but pigmy to the eye. It is extravagant grandiloquence confined to a newspaper about the size of a double letter sheet. He doubtless edits, sets the type, and prints his paper, all alone; but he delights to speak of the concern as if it occupies a block and employs a thousand men.

[Something less than two years ago, Conrad assailed several people mercilessly in his little "People's Tribune," and got himself into trouble. Straightway he airs the affair in the "Territorial Enterprise," in a com-munication over his own signature, and I propose to reproduce it here, in all its native simplicity and more than human candor. Long as it is, it is well worth reading, for it is the richest specimen of journalistic literature the history of America can furnish, perhaps:]

(352)

From the Territorial Enterprise, Jan. 20, 1870.

A SEEMING PLOT FOR ASSASSINATION MISCARRIED.

To the Editor of the Enterprise: Months ago, when Mr. Sutro incidentally exposed mining management on the Comstock, and among others roused me to protest against its continuance, in great kindness you warned me that any attempt by publications, by public meetings, and by legislative action, aimed at the correction of chronic mining evils in Storey County, must entail upon me (*a*) business ruin, (*b*) the burden of all its costs, (*c*) personal violence, and if my purpose were persisted in, then (*d*) assassination, and after all nothing would be effected.

YOUR PROPHECY FULFILLING.

In large part at least your prophecies have been fulfilled, for (*a*) assaying, which was well attended to in the Gold Hill Assay Office (of which I am superintendent), in consequence of my publications, has been taken elsewhere, so the President of one of the companies assures me. With no reason *assigned*, other work has been taken away. With but one or two important exceptions, our assay business now consists simply of the *gleanings* of the vicinity. (*b*) Though my own personal donations to the People's Tribune Association have already exceeded $1,500, outside of our own numbers we have received (in money) less than $300 as contributions and subscriptions for the journal. (*c*) On Thursday last, on the main street in Gold Hill, near noon, with neither warning nor cause assigned, by a powerful blow I was felled to the ground, and while down I was kicked by a man who it would seem had been led to *believe* that I had spoken derogatorily of him. By whom he was so induced to believe I am as yet unable to say. On Saturday last I was again assailed and beaten by a man who first informed me why he did so, and who persisted in making his assault even after the erroneous impression under which he *also* was at first laboring had been clearly and repeatedly pointed out. This same man, after failing through intimidation to elicit from me the names of our editorial contributors, against giving which he knew me to be pledged, beat himself weary upon me with a raw hide, I not resisting, and then pantingly threatened me with permanent disfiguring mayhem, if ever again I should introduce his name into print, and who but a few minutes before his attack upon me assured me that the only reason I was "permitted"

23**

to reach home alive on Wednesday evening last (at which time the
PEOPLE'S TRIBUNE was issued) was, that he deems me only half-witted,
and be it remembered the very next morning I *was* knocked down and
kicked by a man who seemed to be *prepared* for flight.

[*He sees doom impending:*]

WHEN WILL THE CIRCLE JOIN?

How long before the whole of your prophecy will be fulfilled I cannot say, but under the shadow of so much fulfillment in so short a time,
and with such threats from a man who is one of the most prominent ex-
ponents of the San Francisco mining-ring staring me and this whole
community defiantly in the face and *pointing* to a completion of your
augury, do you blame me for feeling that this communication is the last
I shall ever write for the Press, especially when a sense alike of personal
self-respect, of duty to this money-oppressed and fear-ridden community, and of American fealty to the spirit of true Liberty all command me,
and each more loudly than love of life itself, to declare the name of
that prominent man to be JOHN B. WINTERS, President of the Yellow Jacket Company, a political aspirant and a military General?   The
name of his partially duped accomplice and abettor in this last marvel-
ous assault, is no other than PHILIP LYNCH, Editor and Proprietor
of the Gold Hill *News*.

Despite the insult and wrong heaped upon me by John B. Winters,
on Saturday afternoon, only a glimpse of which I shall be able to afford
your readers, so much do I deplore clinching (by publicity) a serious
mistake of any one, man or woman, committed under natural and not
self-wrought passion, in view of his great apparent excitement at the
time and in view of the almost perfect privacy of the assault, I am far
from sure that I should not have given him space for repentance before
exposing him, were it not that he himself has so far exposed the matter
as to make it the common talk of the town that he has horsewhipped
me.   That fact having been made public, all the facts in connection
need to be also, or silence on my part would seem *more* than singular,
and with many would be proof either that I was conscious of some un-
worthy aim in publishing the article, or else that my "non-combatant"
principles are but a convenient cloak alike of physical and moral
cowardice.   I therefore shall try to present a graphic but truthful picture
of this whole affair, but shall forbear all comments, presuming that the
editors of our own journal, if others do not, will speak freely and

fittingly upon this subject in our next number, whether I shall then be
dead or living, for my death will not stop, though it may suspend, the
publication of the PEOPLE'S TRIBUNE.

[*The " non-combatant" sticks to principle, but takes along a friend
or two of a conveniently different stripe :*]

THE TRAP SET.

On Saturday morning John B. Winters sent verbal word to the Gold
Hill Assay Office that he desired to see me at the Yellow Jacket office.
Though such a request struck me as decidedly cool in view of his own
recent discourtesies to me there alike as a publisher and as a stockholder
in the Yellow Jacket mine, and though it seemed to me more like a
summons than the courteous request by one gentleman to another for a
favor, hoping that some conference with Sharon looking to the better-
ment of mining matters in Nevada might arise from it, I felt strongly
inclined to overlook what *possibly* was simply an oversight in courtesy.
But as then it had only been two days since I had been bruised and
beaten under a hasty and false apprehension of facts, my caution was
somewhat aroused. Moreover I remembered sensitively his contempt-
uousness of manner to me at my last interview in his office. I therefore
felt it needful, if I went at all, to go accompanied by a friend whom he
would not dare to treat with incivility, and whose presence with me
might secure exemption from insult. Accordingly I asked a neighbor
to accompany me.

THE TRAP ALMOST DETECTED.

Although I was not then aware of this fact, it would seem that previous to my request this same neighbor had heard Dr. Zabriskie state
publicly in a saloon, that Mr. Winters had told him he had decided
either to kill or to horsewhip me, but had not finally decided on which.
My neighbor, therefore, felt unwilling to go down with me until he had
first called on Mr. Winters alone. He therefore paid him a visit.
From that interview he assured me that he gathered the impression that
he did not believe I would have any difficulty with Mr. Winters, and
that he (Winters) would call on me at four o'clock in my own office.

MY OWN PRECAUTIONS.

As Sheriff Cummings was in Gold Hill that afternoon, and as I
desired to converse with him about the previous assault, I invited him to
my office, and he came. Although a half-hour had passed beyond four

W.**

o'clock, Mr. Winters had not called, and we both of us began preparing to go home. Just then, Philip Lynch, Publisher of the Gold Hill *News*, came in and said, blandly and cheerily, as if bringing good news:

"Hello, John B. Winters wants to see you."

I replied, "Indeed! Why, he sent me word that he would call on me *here* this afternoon at four o'clock!"

"O, well, it don't do to be too ceremonious just now, he's in my office, and that will do as well — come on in, Winters wants to consult with you alone. He's got something to say to you."

Though slightly uneasy at this change of programme, yet believing that in an *editor's* house I ought to be safe, and anyhow that I would be within hail of the street, I hurriedly, and but partially whispered my dim apprehensions to Mr. Cummings, and asked him if he would not keep near enough to hear my voice in case I should call. He consented to do so while waiting for some other parties, and to come in if he heard my voice or thought I had need of protection.

On reaching the editorial part of the *News* office, which viewed from the street is dark, I did not see Mr. Winters, and again my misgivings arose. Had I paused long enough to consider the case, I should have invited Sheriff Cummings in, but as Lynch went down stairs, he said: "*This* way, Wiegand — it's best to be private," or some such remark.

[I do not desire to strain the reader's fancy hurtfully, and yet it would be a favor to me if he would try to fancy this lamb in battle, or the dueling ground or at the head of a vigilance committee.— M. T.:]

I followed, and *without* Mr. Cummings, and without arms, which I never do or will carry, unless as a soldier in war, or unless I should yet come to feel I must fight a duel, or to join and aid in the ranks of a *necessary* Vigilance Committee. But by following I made a fatal mistake. Following was entering a trap, and whatever animal suffers itself to be *caught* should expect the common fate of a caged rat, as I fear events to come will prove.

Traps commonly are not set for *benevolence*.

[*His body-guard is shut out.*]

THE TRAP INSIDE.

I followed Lynch down stairs. At their foot a door to the-left opened into a small room. From *that* room another door opened into yet *another* room, and once entered I found myself inveigled into what many will ever henceforth regard as a private subterranean Gold Hill den, admirably adapted in proper hands to the purposes of murder, raw or disguised, for from it, with both or even one door closed, when too late, I saw that I *could* not be heard by Sheriff Cummings, and from it, BY VIOLENCE AND BY FORCE, I was prevented from making a peaceable exit, when I thought I saw the studious object of this " consultation " was no other than to compass my killing, *in the presence of Philip Lynch as a witness,* as soon as by insult a proverbially excitable man should be exasperated to the point of assailing Mr. Winters, so that Mr. Lynch, by his conscience and by his well-known tenderness of heart toward the rich and potent would be *compelled* to testify that he saw Gen. John B. Winters kill Conrad Wiegand in "self-defence." But I am going too fast.

OUR HOST.

Mr. Lynch was present during the most of the time (say a little short of an hour), but three times he left the room. His testimony, therefore, would be available only as to the bulk of what transpired. On entering this carpeted den I was invited to a seat near one corner of the room. Mr. Lynch took a seat near the window. J. B. Winters sat (at first) near the door, and began his remarks essentially as follows:

" I have come here to exact of you a retraction, in black and white, of those damnably false charges which you have preferred against me in that —— —— infamous lying sheet of yours, and you must declare yourself their author, that you published them knowing them to be false, and that your motives were malicious."

" Hold, Mr. Winters. Your language is insulting and your demand an enormity. I trust I was not invited here either to be insulted or coerced. I supposed myself here by invitation of Mr. Lynch, at your request."

" Nor did I come here to insult you. I have already told you that I am here for a very different purpose."

" Yet your language *has* been offensive, and even now shows strong excitement. If insult is repeated I shall either leave the room or call in Sheriff Cummings, whom I just left standing and waiting for me outside the door."

" No, you won't, sir. You may just as well understand it at once as not. *Here* you are my man, and I'll tell you why ! Months ago you put your property out of your hands, boasting that you did so to escape losing it on prosecution for libel."

"It is true that I did convert all my immovable property into personal property, such as I could trust safely to others, and chiefly to escape ruin through possible libel suits."

"Very good, sir. Having placed yourself beyond the pale of the law, *may God help your soul if you* DON'T make precisely such a retraction as I have demanded. I've got you now, and by —— before you can get out of this room you've *got* to both write and sign precisely the retraction I have demanded, and before you go, anyhow — you —— —— low-lived —— lying —— —— I'll teach you what *personal* responsibility is *outside* of the law; and, by ——, Sheriff Cummings and all the friends you've got in the world besides, can't save you, you —— ——, etc.! *No*, sir. I'm *alone* now, and I'm *prepared* to be shot down just here and now rather than be vilified by you as I have been, and suffer you to escape me after publishing those charges, not only here where I am known and universally respected, but where I am *not* personally known and may be injured."

I confess this speech, with its terrible and but too plainly *implied* threat of killing me if I did not sign the paper he demanded, terrified me, especially as I saw he was working himself up to the highest possible pitch of passion, and instinct told me that any reply other than one of seeming concession to his demands would only be fuel to a raging fire, so I replied:

"Well, if I've *got* to sign ——," and then I paused some time. Resuming, I said, "But, Mr. Winters, you are greatly excited. Besides, I see you are laboring under a total misapprehension. It is your duty not to inflame but to calm yourself. I am prepared to show you, if you will only point out the article that you allude to, that *you* regard as ' charges ' what no calm and logical mind has any *right* to regard as such. *Show* me the charges, and I will try, at all events; and if it becomes plain that no charges *have* been preferred, then plainly there can be nothing to retract, and no one could rightly *urge* you to demand a retraction. You should beware of making so serious a mistake, for however *honest* a man may be, every one is liable to misapprehend. Besides you *assume* that *I* am the author of some certain article which you have not pointed out. It is *hasty* to do so."

He then pointed to some numbered paragraphs in a Tribune article, headed " What's the Matter with Yellow Jacket? " saying " *That's* what I refer to."

To gain time for general reflection and resolution, I took up the paper and looked it over for awhile, he remaining silent, and as I hoped, cooling. I then resumed, saying, "As I supposed. I do not *admit* having written that article, nor have you any right to *assume* so important a point, and then base important action upon your assumption. You might deeply regret it afterwards. In my published Address to the People, I notified the world that no information as to the authorship of any article would be given without the consent of the writer. I therefore cannot honorably tell you *who* wrote that article, nor can you exact it."

"If you are *not* the author, then I *do* demand to know who is?"

"I must decline to say."

"Then, by ——, I brand *you* as its author, and shall treat you accordingly."

"Passing that point, the most important misapprehension which I notice is, that you regard them as 'charges' at all, when their context, both at their beginning and end, show they are not. These words introduce them: '*Such an investigation* [just before indicated], *we think MIGHT result in showing some of the following points.*' Then follow eleven specifications, and the succeeding paragraph shows that the suggested investigation 'might EXONERATE those who are generally believed guilty.' You see, therefore, the context *proves* they are not preferred *as* charges, and this you seem to have overlooked."

While making those comments, Mr. Winters frequently interrupted me in such a way as to convince me that he was *resolved* not to consider candidly the thoughts contained in my words. He insisted upon it that they *were* charges, and "By ——," he would make me take them back *as* charges, and he referred the question to Philip Lynch, to whom I then appealed as a literary man, as a logician, and as an editor, calling his attention especially to the introductory paragraph just before quoted.

He replied, "If they are *not* charges, they certainly are *insinuations*," whereupon Mr. Winters renewed his demands for retraction precisely such as he had before named, except that he would allow me to state who *did* write the article if I did not myself, and this time shaking his fist in my face with more cursings and epithets.

When he threatened me with his clenched fist, instinctively I tried to rise from my chair, but Winters then forcibly thrust me down, as he did every other time (at least seven or eight), when under similar imminent danger of bruising by his fist (or for aught I could know worse than that after the first stunning blow), which he could easily and safely to himself have dealt me so long as he kept me down and stood over me.

This fact it was, which more than anything else, convinced me that by plan and plot I was purposely made powerless in Mr. Winters' hands, and that he did not mean to allow me that advantage of being afoot, which he possessed. Moreover, I then became convinced, that Philip Lynch (and for what *reason* I wondered) would do absolutely nothing to protect me in his own house. I realized then the situation thoroughly. I had found it equally vain to protest or argue, and I would make no unmanly appeal for pity, still less apologize. Yet my life had been by the plainest possible implication threatened. I was a weak man. I was unarmed. I was helplessly down, and Winters was afoot and probably armed. Lynch was the only "witness." The statements demanded, if given and not explained, would utterly sink me in my own self-respect, in my family's eyes, and in the eyes of the community. On the other hand, should I give the author's name how could I ever expect that confidence of the People which I should no longer deserve, and how much dearer to me and to my family was *my* life than the life of the real author to *his* friends. Yet life seemed dear and each minute that remained seemed precious, if not solemn. I sincerely trust that neither you nor any of your readers, and especially none with families, may ever be placed in such seeming *direct* proximity to death while obliged to decide the one question I was compelled to, viz.: What should I do — I, a man of family, and *not* as Mr. Winters is, "alone."

[*The reader is requested not to skip the following.*— M. T.:]

STRATEGY AND MESMERISM.

To gain time for further reflection, and hoping that by a *seeming* acquiescence I might regain my personal liberty, at least till I could give an alarm, or take advantage of some momentary inadvertence of Winters, and then without a *cowardly* flight escape, I resolved to write a certain kind of retraction, but previously had inwardly decided

First.— That I would studiously avoid every action which might be *construed* into the drawing of a weapon, even by a self-infuriated man, no matter what amount of insult might be heaped upon me, for it seemed to me that this great excess of compound profanity, foulness and epithet must be more than a mere indulgence, and therefore must have some object. "Surely in vain the net is spread in the sight of any bird." Therefore, as before without thought, I thereafter by intent kept my hands away from my pockets, and generally in sight and spread upon my knees.

Second.— I resolved to make no motion with my arms or hands which could possibly be construed into aggression.

Third.— I resolved completely to govern my outward manner and suppress indignation. To do this, I must govern my spirit. To do that, by force of imagination I was obliged like actors on the boards to resolve myself into an unnatural mental state and see all things through the eyes of an assumed *character*.

Fourth.— I resolved to try on Winters, silently, and unconsciously to himself a mesmeric power which I possess over certain kinds of people, and which at times I have found to work even in the dark over the lower animals.

Does any one smile at these last counts? God save you from ever being *obliged* to beat in a game of chess, whose stake is your life, you having but four poor pawns and pieces and your adversary with his full force unshorn. But if you are, provided you have any strength with breadth of will, do not despair. Though mesmeric power may not *save* you, it may help you; *try* it at all events. In this instance I was conscious of power coming into me, and by a law of nature, I know Winters was correspondingly weakened. If I could have gained more time I am sure he would not even have struck me.

It takes time both to form such resolutions and to recite them. That time, however, I gained while thinking of my retraction, which I first wrote in pencil, altering it from time to time till I got it to suit me, my aim being to make it look like a concession to demands, while in fact it should tersely speak the truth into Mr. Winters' mind. When it was finished, I copied it in ink, and if correctly copied from my first draft it should read as follows. In copying I do not think I made any material change.

COPY.

To Philip Lynch, Editor of the Gold Hill News : I learn that Gen. John B. Winters believes the following (pasted on) clipping from the PEOPLE'S TRIBUNE of January to contain distinct charges of mine against him personally, and that as such he desires me to retract them unqualifiedly.

In compliance with his request, permit me to say that, although Mr.

Winters and I see this matter differently, in view of his strong feelings in the premises, I hereby declare that I do not know those "charges" (if such they are) to be true, and I hope that a critical examination would altogether disprove them. CONRAD WIEGAND.

Gold Hill, January 15, 1870.

I then read what I had written and handed it to Mr. Lynch, where-upon Mr. Winters said:

"That's not satisfactory, and it won't do;" and then addressing himself to Mr. Lynch, he further said: "How does it strike *you* ?"

"Well, I confess I don't see that it *retracts* anything."

"Nor do I," said Winters; "in fact, I regard it as adding insult to injury. Mr. Wiegand, you've got to do better than that. *You* are not the man who can pull wool over *my* eyes."

"That, sir, is the only retraction I can write."

"No it isn't, sir, and if you so much as *say* so again you do it at your peril, for I'll thrash you to within an inch of your life, and, by ——, sir, I don't pledge myself to spare you even that inch either. I want you to understand I have asked you for a very different paper, and that paper you've got to sign."

"Mr. Winters, I assure you that I *do* not wish to irritate you, but, at the same time, it is utterly *impossible* for me to write any other paper than that which I have written. If you are resolved to *compel* me to *sign* something, Philip Lynch's hand must write at your dictation, and if, when written, I *can* sign it I will do so, but such a document as you say you *must* have from me, I never can sign. I mean what I say."

"Well, sir, what's to be done must be done quickly, for I've been here long enough already. I'll put the thing in another shape (and then pointing to the paper); don't you know those charges to be false?"

"I do not."

"Do you know them to be true?"

"Of my own personal knowledge I do not."

"Why then did you print them?"

"Because rightly considered in their connection they are *not* charges, but pertinent and useful *suggestions* in answer to the queries of a corre-spondent who stated facts which are inexplicable."

"Don't you know that *I* know they are false?"

"If you *do*, the proper course is simply to deny them and court an investigation."

"And do YOU claim the right to make ME come out and deny
anything you may choose to write and print?"

To that question I think I made no reply, and he then further said:
"Come, now, we've talked about the matter long enough. I want your
final answer — did you write that article or not?'

"I cannot in honor tell you *who* wrote it."

"Did you not see it before it was printed?"

"Most certainly, sir."

"And did you deem it a fit thing to publish?"

"Most assuredly, sir, or I would never have consented to its appearance. Of its *authorship* I can say nothing whatever, but for its *publication* I assume full, sole and personal responsibility."

"And do you then retract it or not?"

"Mr. Winters, if my refusal to sign such a paper as you have
demanded *must* entail upon me all that your language in this room
fairly implies, then I ask a few minutes for prayer."

"Prayer! —— —— you, this is not your *hour* for prayer — your
time to pray was when you were writing those —— lying charges. Will
you sign or not?"

"You already have my answer."

"What! do you still refuse?"

"I do, sir."

"Take *that*, then," and to my amazement and inexpressible relief
he drew only a rawhide instead of what I expected — a bludgeon or
pistol. With it, as he spoke, he struck at my left ear downwards, as if
to tear it off, and afterwards on the side of the head. As he moved
away to get a better chance for a more effective shot, for the first time
I gained a chance under peril to rise, and I did so pitying him from the
very bottom of my soul, to think that one so naturally capable of true
dignity, power, and nobility could, by the temptations of this State, and
by unfortunate associations and aspirations, be so deeply debased as to
find in such brutality anything which he could call satisfaction — but the
great hope for us all is in progress and growth, and John B. Winters, I
trust, will yet be able to comprehend my feelings.

He continued to beat me with all his great force, until absolutely
weary, exhausted, and panting for breath. I still adhered to my purpose of non-aggressive defence, and made no other use of my arms than
to defend my head and face from further disfigurement. The mere
pain arising from the blows he inflicted upon my person was of course

transient, and my clothing to some extent deadened its severity, as it now hides all remaining traces.

When I supposed he was through, taking the butt end of his weapon and shaking it in my face, he warned me, if I correctly understood him, of more yet to come, and furthermore said, if ever I again dared introduce his name to print, in either my own or any other public journal, he would cut off my left ear (and I do not *think* he was jesting) and send me home to my family a visibly mutilated man, to be a standing warning to all low-lived puppies who seek to blackmail gentlemen and to injure their good names. And when he *did* so operate, he informed me that his implement would not be a whip but a knife.

When he had said this, unaccompanied by Mr. Lynch, as I remember it, he left the room, for I sat down by Mr. Lynch, exclaiming: "The man is mad — he is *utterly* mad — this step is his ruin — it is a mistake — it would be ungenerous in me, despite of all the ill usage I have here received, to expose him, at least until he has had an opportunity to reflect upon the matter. I shall be in no haste."

"Winters *is* very mad just now," replied Mr. Lynch, "but when he is himself he is one of the finest men I ever met. In fact, he told me the reason he did not meet you upstairs was to spare you the humiliation of a beating in the sight of others."

I submit that that unguarded remark of Philip Lynch convicts him of having been privy in advance to Mr. Winters' intentions whatever they may have been, or at least to his meaning to make an assault upon me, but I leave to others to determine how much censure an *editor* deserves for inveigling a weak, non-combatant man, also a publisher, to a pen of his own to be horsewhipped, if no worse, for the simple printing of what is verbally in the mouth of nine out of ten men, and women too, upon the street.

While writing this account two theories have occurred to me as *possibly* true respecting this most remarkable assault:

First — The aim *may* have been simply to extort from me such admissions as in the hands of money and influence would have sent me to the Penitentiary for libel. This, however, seems unlikely, because any statements elicited by fear or force could not be evidence in law or could be so explained as to have no force. The statements wanted so badly must have been desired for some other purpose.

Second — The other theory has so dark and willfully murderous a look that I shrink from writing it, yet as in all probability my death at the earliest practicable moment has already been decreed, I feel I should do all I can before my hour arrives, at least to show others how to break up that aristocratic rule and combination which has robbed all Nevada of true freedom, if not of manhood itself. Although I do not prefer this hypothesis as a "*charge*," I feel that as an American citizen I still have a right both to think and to speak my thoughts even in the land of Sharon and Winters, and as much so respecting the theory of a brutal assault (especially when I have been its subject) as respecting any other apparent enormity. I give the matter simply as a suggestion which may explain to the proper authorities and to the people whom they should represent, a well-ascertained but notwithstanding a darkly mysterious fact. The scheme of the assault *may* have been

First — To terrify me by making me conscious of my own helplessness after making actual though not legal threats against my life.

Second — To imply that I could save my life only by writing or signing certain specific statements which if not subsequently explained would eternally have branded me as infamous and would have consigned my family to shame and want, and to the dreadful compassion and patronage of the rich.

Third — To blow my brains out *the moment I had signed*, thereby preventing me from making any subsequent explanation such as *could* remove the infamy.

Fourth — Philip Lynch to be compelled to testify that I was killed by John B. Winters in self-defence, for the conviction of Winters would bring *him* in as an accomplice. If that *was* the programme in John B. Winters' mind nothing saved my life but my persistent *refusal* to sign, when that refusal seemed clearly to me to be the choice of death.

The remarkable assertion made to me by Mr. Winters, that pity only spared my life on Wednesday evening last, almost compels me to believe that at first he *could* not have intended me to leave that room alive; and why I was allowed to, unless through mesmeric *or some other invisible influence*, I cannot divine. The more I reflect upon this matter, the more probable as true does this horrible interpretation become.

The narration of these things I might have spared both to Mr. Winters and to the public had he himself observed silence, but as he has both verbally spoken and suffered a thoroughly garbled statement of facts to appear in the Gold Hill *News* I feel it due to myself no less than to this community, and to the entire independent press of America and Great Britain, to give a true account of what even the Gold Hill *News* has pronounced a disgraceful affair, and which it deeply regrets because of some alleged telegraphic mistake in the account of it. [Who received the erroneous telegrams?]

Though he may not deem it prudent to take my life just now, the publication of this article I feel sure must compel Gen. Winters (with his peculiar views about *his* right to exemption from criticism by *me*) to resolve on my violent death, though it may take years to compass it. Notwithstanding *I* bear *him* no ill will; and if W. C. Ralston and William Sharon, and other members of the San Francisco mining and milling Ring feel that he above all other men in this State and California is the most fitting man to supervise and control Yellow Jacket matters, until I am able to vote more than half their stock I presume he will be retained to grace his present post.

Meantime, I cordially invite all who know of any sort of important villainy which only *can* be cured by exposure (and who would expose it if they felt sure they would not be betrayed under bullying threats), to communicate with the PEOPLE'S TRIBUNE; for until I *am* murdered, so long as I can raise the means to publish, I propose to continue my *efforts* at least to revive the liberties of the State, to curb oppression, and to benefit man's world and God's earth.

                            CONRAD WIEGAND.

[It does seem a pity that the Sheriff was shut out, since the good sense of a general of militia and of a prominent editor failed to teach them that the merited castigation of this weak, half-witted child was a thing that ought to have been done in the street, where the poor thing could have a chance to run. When a journalist maligns a citizen, or attacks his good name on hearsay evidence, he deserves to be thrashed for it, even if he *is* a " non-combatant " weakling; but a generous adversary would at least allow such a lamb the use of his legs at such a time.— M. T.]